Perspectives on Ottawa's High-tech Sector

P.I.E. Peter Lang

Bruxelles · Bern · Berlin · Frankfurt am Main · New York · Oxford · Wien

Canadian Studies

The series Canadian Studies examines the many facets of Canadian reality from a multidisciplinary perspective. Contributions from both the humanities and the social sciences are invited. The editor welcomes manuscripts whose primary object is "Canada" in the widest possible sense of the term. The series therefore covers a variety of fields such as literature, history, sociology, politics, economics, geography, law, media, museology, etc. as well as comparative studies.

One of the most innovative features of this series is its focus on the latest research conducted outside Canada. It therefore illuminates various aspects of the country in a new and significant manner and encourages a constant and innovative dialogue between Canadian scholars and the community of Canadian studies specialists worldwide.

Series editor: **Serge Jaumain**

Centre d'études canadiennes

Université Libre de Bruxelles (Belgique)

Nick NOVAKOWSKI and Rémy TREMBLAY (eds.)

Perspectives on Ottawa's High-tech Sector

Canadian Studies
No. 13

Éditions scientifiques internationales
Brussels, 2007
1 avenue Maurice, B-1050 Brussels, Belgium
info@peterlang.com; www.peterlang.com

ISSN 1781-3867
ISBN 978-90-5201-370-1
D/2007/5678/57

Bibliographic information published by "Die Deutsche Bibliothek"
"Die Deutsche Bibliothek" lists this publication in the "Deutsche Nationalbibliografie"; detailed bibliographic data is available in the Internet at <http://dnb.ddb.de>.

CIP available from the British Library, GB and the Library of Congress, USA.

Contents

Preface 9
Nick Novakowski and Rémy Tremblay

INTRODUCTION. Perspectives on Ottawa's High-tech Sector. Ottawa in the Age of the New Economy 11
Nick Novakowski

PART I. OTTAWA: A KNOWLEDGE CITY?

CHAPTER 1. Ottawa – Canada's Capital. The Emergent Knowledge City 23
Nick Novakowski and Rémy Tremblay

CHAPTER 2. Ottawa – North American Technopole. Fact or Fiction? 43
Rémy Tremblay

PART II. PLANNING THE CLUSTER. BY DECISION, BY DESIGN OR BY DESTINY?

CHAPTER 3. Local Governments' Record of Assessing the Impacts of the High-tech Industry on Ottawa's Land Use-Transportation Relationship. 1970s-2005 61
Barry Wellar and Nick Novakowski

CHAPTER 4. Unanticipated Benefits. The Role of Planning in the Development of the Ottawa Region Technology Industries 91
David Gordon and Betsy Donald, with John Kozuskanich

Part III. 'Growing' the Cluster. Idea Farming and Innovation Strategies for Economic Development

Chapter 5. Steering on Black Ice. The Continuing Search for Sustainable Livelihoods in the Ottawa Tech Sector ... 119
Edward T. Jackson and Rahil Khan

Chapter 6. The Renaissance of Ottawa's ICT Community Through Joint Initiatives ... 145
David Large

Chapter 7. Total Creative Revolution. Innovation and Governance at KAO Infosystems Canada ... 157
Christopher Wilson

Chapter 8. Comparing Science and Engineering Employment in Canadian and US Cities ... 173
Desmond Beckstead and W. Mark Brown

Chapter 9. Reflexive Governance and City-Region Economic Change. Between Reflexivity and State Selectivity in the Governance of Ottawa's Economic Transition ... 199
Joseph Leibovitz

Part IV. The Unique Ottawa Cluster. Regional, Bilingual and Cosmopolitan

Chapter 10. A Cross-Border High-tech Cluster. A View From Gatineau ... 225
Guy Chiasson

Chapter 11. Immigrants in Ottawa's Information Technology Sector. A Profile ... 245
Amelia Brown

Conclusion ... 263
Hugues Chicoine and Rémy Tremblay

Contributors ... 267

Preface

The present book was inspired by the emergence of one of the newest concepts in urban growth, development and planning: the knowledge city. Academic interest in the topic is developing because the cities of *the core* need new knowledge products and services in order to grow and/or remain competitive. In this era of manufacturing and service jobs that migrate to *the periphery*, there must be a new vision of the city. The question is: how does Ottawa figure into the discussion? The high-tech sector in the Ottawa area was nascent in the 1950s and even earlier. And, the city blossomed under the high-tech surge of the late 1990s. But, with the downturn in 2001, many unanswered questions appeared. In this book, we attempt to address some of those questions.

Recently, the City of Ottawa has become aware of the work of other forward-looking municipalities and has actively engaged in planning for the high-sector sector (e.g., look no further than the Ottawa 20/20 Broadband Plan). However, this planning vision is new (and not to mention late) as is demonstrated in a number of the chapters in this book. The challenge of the book was to seek out a variety of points of view on the high-tech sector, and to wrestle with the realisation that Ottawa and its region is in a transitional phase.

This book is the result of the work of many people, including the many contributing authors who were both patient and quick with their responses to queries from the editors. Over and above acknowledgements recorded in individual chapters, we would also like to thank Sandra Kuzniak and Serge Jaumain for their guidance from the Peter Lang office in Brussels.

Finally, the support of the Canada Research Chair Council was crucial and much appreciated.

Nick Novakowski
Memorial University of Newfoundland – Grenfell College

Rémy Tremblay
Université du Québec à Montréal

August 22, 2007

Preface

[illegible]

INTRODUCTION

Perspectives on Ottawa's High-tech Sector

Ottawa in the Age of the New Economy

Nick NOVAKOWSKI

Memorial University of Newfoundland

While Ottawa is the national capital of Canada, it can also be perceived as a statistical jurisdiction defined by Statistics Canada as a Census Metropolitan Area (CMA). As such, it is then known as Ottawa-Gatineau. Bridging two provinces, Ontario and Québec, Ottawa-Gatineau is North America's most truly bilingual city. In fact, the University of Ottawa is the continent's only bilingual university, where the curriculum is offered in both English and French.

The first inhabitants of the region were Algonquins, who called the Ottawa River the Kichesippi or Great River. They referred to themselves as the Kichesippirini or People of the Great River. Although their archaeological importance is often eclipsed by Ottawa's current role as the national capital, their legacy remains in the burial grounds and relics that have been maintained by Culture, Heritage and Recreation Canada; Collections Canada; the National Capital Commission and other federal, provincial and local agencies.

The city takes its name from the Outaouais, a group of trading middlemen that were only temporary residents in the area, but made contact with the original French explorers of the modern era after the Iroquois had driven the Algonquin from the region. The Ottawa area came under French rule first, and then later British rule. In 1867, Canadian independence from the yoke of its colonial past jumpstarted Ottawa into the national consciousness. Ottawa, because of its advantageous geographical location as a compromise between Upper and Lower Canada, became a central component of the national story.

The city's official motto, 'Advance,' was military in origin in 1882, but is now more appropriate for capturing the city's knowledge-based growth. In fact, almost as many people now work in the *knowledge economy* as they do for the federal government, and the region has experienced a significant increase in population and new companies.

According to Hall (2000) in his typology of capital cities, Ottawa is a political capital, which means that it was 'created' as a seat of government (as opposed to Toronto, Montréal, Québec City or Kingston) and lacks the commercial and culture-making origins of other, older cities seen in a particular nation. Ottawa is a federal capital like Bern, Brasilia, Canberra, Moscow, or Washington, DC. But, all of these federal capitals have their own question: what dominates – the vernacular city or the federal city? With respect to Ottawa, the debate is open. It was not pre-planned in the way that Brasilia, Canberra or Islamabad were. Nor does the federal city dominate in the way that Washington's does. So, perhaps Ottawa is both. It is a government city while its private sector simultaneously flourishes.

This book developed out of a need to address the new Ottawa, the globalising Ottawa. In more developed countries, wealth-generation through product manufacturing is saturated as wage levels are high and production processes cannot compete with the advantages accruing to the international division of labour (Bakan, 2004). In its place, new types of cities are emerging that create and nurture their own comparative advantages. Some of these cities are network cities like the Randstad in the Netherlands, or megalopolitan structures like Boston-Washington or Tokyo-Nagoya.

Another specific type of new city is *the knowledge city*. This new type of city is both the beneficiary of historical trends that support the emergence of knowledge centres and is actively planned for. The only way to tap into true wealth creation in the developed world is through knowledge creation and innovation. Setting the groundwork for innovation are the characteristics of the New Economy – an economy that is based on learning, education, invention and innovation. Chapter 1 of this collection will present the attributes of both the knowledge city and the New Economy.

There is little doubt that we live in a knowledge era where socio-economic hierarchies are repositioning themselves while being simultaneously dependent on the ownership of information rather than the factors of production. The new unit of analysis is innovation. The knowledge city is a city that is "purposefully designed to encourage the nurturing of knowledge" (Edvinsson *et al.*, 2004). Innovation, therefore, is the basis of the knowledge city nexus – and the point of departure of this book.

Readers can see innovation and the elements of the New Economy revealed by the different authors detailing different accounts and analyses of the Ottawa area. The book is organised into four themes:

1. Ottawa: A Knowledge City?
2. Planning the Cluster: By Decision, By Design or By Destiny?
3. 'Growing' the Cluster: Idea Farming and Innovation Strategies for Economic Development, and
4. The Unique Ottawa Cluster: Regional, Bilingual and Cosmopolitan.

The authors of the chapters have different analyses to present. In sequence, their topics have the following themes. In Chapter 1, Nick Novakowski and Rémy Tremblay introduce the Ottawa area and its fundamental characteristics as a knowledge city. They employ statistics and field observations to provide insight into Ottawa's economic profile, its growth and some of its unique characteristics. In Chapter 2, Tremblay introduces the idea of how local businesses perceive themselves – whether as a part of a realistic high-tech cluster or as a secondary cluster. He uses a survey to quantify his investigation and makes a number of startling observations about how Ottawa ranks as a high-tech centre. Every year, dozens of rankings are published on different aspects of American and Canadian cities. However, few of them propose integrated rankings on a North American scale since cross-comparative data of both countries is limited. In this chapter, Tremblay examines rankings of Canadian and American cities of more than 500,000 people based on indicators related to human capital and high-technology employment. Through an integrated analysis of Canadian and American cities, he discovers that the Ottawa area is competitive with several high-tech centres south of the border.

In Chapter 3, the planning of the cluster as a theme is explored. Here, Barry Wellar and Novakowski assess the impact of the high-tech industry on Ottawa's land use-transportation system relationship. In the interests of space, and the availability of documentation on the past, present and future of the land use-transportation system relationship, the study time period is confined to the last 35 years and the next 10 years, that is, 1970-2015. And, for the same reasons, the area of geographic coverage that is referred to as Ottawa involves the Ontario side of the National Capital Region (NCR).

Two general findings about the land use-transportation system relationship emerge from the impacts study. First, up until the late 1970s the impact of the high-tech industry on the relationship was relatively minor, and largely uneventful. The amount and rate of industry growth was such that it could be accommodated with only modest changes to either land use or transportation policies, plans or programmes. At the same time, signals were coming from politicians, land speculators and developers (and their agents) that the high-tech industry could be presented in due course as the private sector catalyst to move Ottawa out from under the long shadow cast by the federal government as the NCR's dominant employer, landlord, tenant, trip generator, and travel destination.

By the mid-1980s, Ottawa or 'Silicon Valley North' began to experience two significant changes in the land use-transportation system relationship as a result of rapid, large expansions to the high-tech industry. The initial set of changes involved sharp increases in the requests for and approvals of official plan amendments and rezoning applications at both the municipal and regional government levels for lands located primarily in the western sector (Kanata and West Carleton) of Ottawa. And, as an associated consequence, there was an increase in road network expansions to accommodate the increased car traffic, again primarily in the western sector but also throughout the NCR due to the widespread distribution of high-tech workers' places of residence.

After examining the impacts of the high-tech industry on Ottawa's land use-transportation system relationship up to 2005, Wellar and Novakowski consider the industry's impacts on the relationship through to 2015. A matter of particular interest is assessing the likely impacts that land use and transportation system decisions taken up to 2005 to serve the high-tech industry can have on Ottawa's smart growth, sustainable development and O-Train strategies over the next decade.

In Chapter 4, David Gordon, Betsy Donald and John Kozuskanich address whether planning actually matters in generating Ottawa's high-tech arena. The chapter opens with a description of Ottawa-Hull (later Ottawa-Gatineau) in the mid-20th century, an industrial urban area still making the transition from lumber to government, and an unattractive location for most knowledge workers. The authors review the 20th century history of urban planning in the region, including the federal government's initial failures to build a "capital worthy of the nation" despite five decades of effort and a number of major plans.

The 1938 and 1950 plans prepared by Jacques Gréber appear to have been quite influential in transforming the region from a lumber town into a metropolis that was far more attractive to civil servants, diplomats, tourists and, quite unintentionally, high-tech employees. The ring of research laboratories established in a greenbelt at the outer edge of the urban area was very much a deliberate planning strategy. It started with the 1942 opening of the National Research Council Montreal Road labs and continued into the mid-1990s expansion of the critical Nortel Networks research campus in the western Greenbelt.

Most second and third generation spin-off companies located in industrial parks initially planned for traditional manufacturing uses by the National Capital Commission and local governments. Local and regional plans from the 1960s to 1980s showed few policy initiatives or even awareness of the industry. Local and regional governments finally understood the importance of the emerging cluster in the mid-1990s and embraced it as a means to diversify and grow the local economy. Some well-regarded community visioning and capacity building was done by the privately-led TOP (The Ottawa Partnership) for the Ottawa 2020 plan, just in time for the 2001 dot.com and telecom meltdown. The authors' tentative conclusion is that planning probably had little effect on the processes that Michael Porter describes regarding the development of clusters, but may have indirectly improved the quality of life valued by the creative class portrayed by Richard Florida.

In Chapter 5, Edward Jackson and Rahil Khan address the managing of disequilibrium. Here, one sees that opportunities for facilitating or growing the Ottawa cluster is the priority. The bursting of the tech bubble in Ottawa undermined the livelihood prospects of many knowledge workers and their families. A wide range of responses ensued: migration to other tech centres in North America or Asia; migration to other sectors, notably lower paying service jobs; transiting to government technology employment; founding new business start-ups; and engagement in re-training. Perhaps the most comprehensive effort in the Ottawa-Gatineau region to navigate amidst the changes of the technology cluster's labour market has been the Ottawa Talent Initiative (OTI). This chapter examines the range of efforts of knowledge workers to manage this volatility and instability, including the OTI. The implications of this body of experience are drawn for the Capital region and for other technology centres.

Chapter 6, meanwhile, involves David Large's investigation of the use of joint initiatives in jump-starting Ottawa's ICT community. The Ottawa ICT community has gone through some rough times over the past five years, but now seems to be on the rebound. However, there have been some painful lessons learned, and market realities that may not have changed command remedial attention. Government, business and civic sector groups must work together to ensure that the ICT renaissance is accompanied by new joint initiatives to help ensure even greater strength and longevity. This chapter suggests a few possible directions for action.

Building on the idea farming in previous chapters, Chapter 7 is Christopher Wilson's case study of KAO Infosystems Canada and how this unique organisation adapted to changes in the high-tech sector. For modern companies, global competition forces an unforgiving regime of continuous change and innovation. The 'new economy' confirms the assertion by the economist Joseph Schumpeter that innovation, rather than price, drives competition. Among computer, information and telecommunications companies, this understanding is common currency, but it is less recognised in basic manufacturing operations. In addition, it is often assumed that innovation is exclusively concerned with new products, ignoring the crucial roles played by the development of new manufacturing processes, of new organisational styles and of new markets. For today's managers, therefore, fostering all forms of innovation must be an all-encompassing task to maintain competitiveness.

Typically, it is also assumed that innovation is the result of an individual genius in a lab or some young whiz kids in their garage. That perspective is complemented by the image of serendipitous discovery – the sudden flash of insight that changes everything. Increasingly, however, innovation is being understood as a social process…the product of many perspectives being brought to bear on an existing problem. The project team or business alliance is *the* industry standard by which firms engage in collective learning. This has been observed in the Cisco versus Nortel fights or the Wintel versus Sun-Netscape-AOL battles, and that the coordination of these collective processes is an issue of major strategic importance. Understandably, the challenge of innovation becomes inevitably linked to the challenges of distributed governance, or how individuals and groups work together, including their establishment of shared vision, effective communication, common understanding, reliable commitment and trust.

Simply put, innovation is dependent on the ability of employees and business partners to construct effective systems of collective learning. Senge (1990) and others have developed a rich understanding of organisational learning in this regard and Moore (1996), Storper (1997) and others have tried to map the nature and competencies for sustaining networked production systems. The purpose of this chapter is to juxtapose these issues of organisational learning and governance side by side through the experience of an extremely successful manufacturing plant, KAO Infosystems of Arnprior, Canada.

During the period covered in the case writing, KAO's diskette and CD products were not subject to particularly radical change, although historically they had been. However, innovative manufacturing processes and organisational structures were required to meet the demands outlined in the case. As such, KAO's experience may have more relevance to managers in traditional industries than the experiences of computer and telecommunications related firms where product innovations tend to dominate.

In the next chapter, Chapter 8, Desmond Beckstead and Mark Brown address innovation capabilities by comparing science and engineering employment in Canadian and American cities. In recent years, cities have become increasingly interested in their ability to generate, attract and retain human capital. One measure of human capital is employment in science and engineering-based occupations. This chapter provides a comparison of the employment shares of these specialised occupations across Canadian and US cities by using data from the Canadian and the US censuses from 1980-81 and 2000-01. The chapter, therefore, provides a perspective on how Canadian cities performed relative to their US counterparts over a twenty year period. It also seeks to evaluate how cities of different sizes have performed because large cities may be advantaged over smaller cities in terms of factors influencing both the demand for and the supply of scientists and engineers.

Chapter 9 is Joseph Leibovitz's contribution to the idea farm for 'growing the cluster' and involves the adaptability of governance and finding new governmental mechanisms. The 1990s saw dramatic economic change in the Ottawa-Gatineau city-region. The city-region, at one time almost entirely dependent on government and related service occupations, experienced massive cuts in public sector employment, while high-technology industries grew rapidly. It was also a time when cultural change and institutional transformation took place in an effort to anticipate, accommodate and plan for this dramatic change. Looking back at this period in perspective, this chapter examines the institutional foundations of this change. Institutions are taken here to mean more than merely organisations, but also a set of conventions, norms, routines and expectations that underwrite the 'rules of the game' in a system of governance.

The dramatic changes in the political economy of Ottawa during the 1990s thus provides an opportunity to analyse how city-region economic governance institutions operated, and in particular to assess the range of activities that took place in light of conceptions related to associative behaviour and reflexivity. Such conceptions have informed the work of economic geographers and economic sociologists who have attempted to make sense of the ability of certain territorial economies (cities and regions) to reassert their advantages even under conditions of globalisation, technological change, increasing competition, and uncertainty. But the chapter also suggests that power relations also play a part in economic governance. As the case of Ottawa demonstrates, there were important elements of conflict and selectivity regarding the types and scale for encouraging certain economic activities at the same time that the institutional support for the high-technology sector grew ever stronger.

The final two chapters of the book address an important component of Ottawa-Gatineau's uniqueness: its sense of both formal and informal regionalism, its bilingualism (official and otherwise), and its cosmopolitan nature. In Chapter 10, Guy Chiasson reflects on the role of regional public institutions in Gatineau. In doing so, he articulates many of the differences that are evident on the two sides of the Ottawa River. At the same time, he is able to identify strengths on the Gatineau side that must be explored. Following in the path set out by the *regional innovation systems* thesis (Wolfe, 2002), this chapter examines the contribution of regional public institutions in the establishment of a high-technology complex in Gatineau. Chiasson looks not only at the ability of the network of regional institutions to initiate technological growth regionally but also its capacity to incorporate social innovation (youth and female entrepreneurship, high-technology development in rural areas, etc.) in the midst of the emerging private new economy.

Chapter 11 is Amelia Brown's statistical investigation of immigration and in-migration to Ottawa's high-tech sector. A significant though often overlooked factor in Ottawa's development into a high-tech centre has been international migration. Nearly 6,000 skilled immigrants in technology occupations came to Ottawa from 1990-2002, representing one third of all skilled immigration to the city. This chapter provides a general overview of trends in the immigration of technology professionals to Ottawa over the past ten years, and focuses on recent Chinese immigration in particular. Since 1993, mainland China has consistently been the largest source of skilled immigration to Ottawa in technology occupations. Drawing on data from Citizenship and Immigration Canada's Landed Immigrants Data System (LIDS) and data from interviews with a small sample of Chinese technology workers in Ottawa, the chapter provides a demographic profile of Chinese immigrants in technology occupations in Ottawa and examines these workers' job and social experiences since the economic downturn in the technology sector. The chapter concludes by addressing the policy implications of these findings at both the local and national levels.

References

Bakan, J. (2004). *The corporation: the pathological pursuit of profit and power*. New York: Simon and Schuster.

Edvinsson, L., Dvir, R., Pasher, E. (2004). Innovations: the new unit of analysis in the knowledge era-the quest and context for innovation efficiency and management of IC. *Journal of Intellectual Capital*, *5* (1), 40-58.

Hall, P. (2000). The Changing Role of Capital Cities. *Plan Canada*, *40* (3), 8-11.

Moore, J.F. (1996). *The death of competition - leadership and strategy in the age of business ecosystems*. New York: HarperCollins.

Senge, P. (1990). *The fifth discipline*. New York: Doubleday.

Storper, M. (1997). *The regional world: territorial development in a global economy*. New York: Guilford Press.

Part I

Ottawa: A Knowledge City?

CHAPTER 1

Ottawa – Canada's Capital

The Emergent Knowledge City

Nick NOVAKOWSKI and Rémy TREMBLAY

Memorial University of Newfoundland
and Université du Québec à Montréal

Introduction

Canada's capital – Ottawa – is earning a reputation as a global technology centre that offers a dynamic mix of economic, cultural, educational and recreational opportunities. Almost as many people now work in the *knowledge economy* as in government and the region has experienced a significant increase in population and new companies, the result of strategic multi-stakeholder initiatives. Today, Ottawa is an advanced technology centre, particularly known for its research and development in the fields of telecommunications, technology services, software development, defence and security, microelectronics/wireless and photonics. In this chapter, we employ the characteristics of *the knowledge city* and *the New Economy* in order to explore the dramatic reorientation of Ottawa's economy over the past 50 years. Overall, the successful knowledge city makes the transition regarding its development goals from *hunting and gathering* (i.e., industrial recruitment through tax breaks, infrastructure development, financial incentives) to *gardening* (i.e., promoting growth from within a cluster through education, research and development, and innovation). Innovative capacity is the foundation of all knowledge cities.

The Knowledge City

A specific type of new city is *the knowledge city*, one that is both the beneficiary of historical trends that support the emergence of knowledge centres and is actively planned for. Since international investment originating in more developed countries (MDCs) continues to flow primarily from MDC to MDC, cities within the global hierarchy need a new way in which to differentiate themselves and remain competitive. Currently, the strongest way to remain competitive is in terms of knowledge creation and innovation.

The knowledge city is sometimes referred to by a variety of other names: the ideopolis (Work Foundation, 2006); the technopole (Tremblay, 2004a), the intelligent city (Komninos, 2002); and the cyborg city (Picon, 1998). Beginning with what the knowledge city is not, it is not a network city (although it may be a part of one like Delft). Within the urban studies and planning literature, the network city is a polynucleated city where two or more cities are highly connected by communication and transportation linkages (Batten, 1995). Nor is the knowledge city a digital city, although the digital city is certainly one dimension of the knowledge city. Rather, the digital city refers to the local social information infrastructure that serves local residents and visitors to a city in terms of their social and cultural needs (e.g., entertainment, education, legal and government services, etc.). In another sense, the digital city refers to a free space to experience and experiment with cyberspace (van den Besselaar, Melis and Beckers, 2000). The knowledge city can have components of the network and digital cities, but is much more.

The term *knowledge city* remains an *essentially contested concept*, which means that it is appraisive and internally complex. At the same time, there are certain characteristics of the knowledge city that seem to resonate in the literature and provide a point of departure for applications. The true knowledge city is more than an inventory of cultural, economic and technological parameters. Rather, the knowledge city is a complex constellation of cultural and economic factors that combine to facilitate innovation. We live in the information and knowledge era where the new unit of analysis is innovation (Edvinsson *et al.*, 2004). The knowledge city, therefore, is a city that is "purposefully designed to encourage the nurturing of knowledge" (Edvinsson *et al.*, 2004).

Knowledge cities are a nexus of universities, international airports, cosmopolitan realities, and research and development. They are characterised by high per capita levels of scientists and engineers, venture capitalists, patent generation, and advanced technology jobs. They are not former industrial cities that are seeking an ephemeral new license to attract investment and tourism. Rather, they are new hybrids benefiting from the inertia of their universities and international connections and umbilici to learning, government, funding, and banking services.

Knowledge cities demonstrate many elements of the New Economy, but reach beyond it. The framework of analysis used for measuring the presence of the New Economy by the Progressive Policy Institute involves at least some focus on high-technology manufacturing. Yet, the knowledge city is about far more than product manufacturing. Manufacturing a silicon chip is, after all, quite different and not as interesting as learning about or designing a better one. The knowledge city demonstrates a complex variety of innovation engines, which include (among other possibilities) the following: universities, big events, libraries, museums, the gate, the future outlook towers, capital markets, digital infrastructure, post-industrial science parks, brownfield industrial zones ripe for infill, urban virtual space, and perhaps surprisingly, coffee shops or cafés and other so-called third places (after home and work) like malls (Dvir and Pasher, 2004; Ergazakis, Metaxiotis and Psarras, 2004; El-Diraby and Wang, 2005).

Simultaneous to the emergence of the knowledge city, we are experiencing another massive economic reorganisation with respect to the leisure economy's operation within the global economy. Authors like Molitor (2005) suggest that leisure activities (entertainment, tourism, sports, events, and education) are soon be the world's dominant employer. According to Moliter (2000), the global economy is well on its way into restructuring itself into the so-called *leisure economy*, which is to become the globe's dominant employer by 2015. With the leisure era, the focus is less on materialism and more on experiences. Services for education, entertainment, travel, sports and recreation are to come to dominate global employment within the decade. The workweek is to contract, and available free time is to increase. Particularly since new technologies can deliver new leisure experiences, these two mega-trends of knowledge services and leisure services must be considered in tandem as the overlap is considerable. In this chapter, we investigate Ottawa, and Ottawa-Gatineau as a manifestation of the New Economy, and then elaborate on the city's potential for growth beyond that platform.

Introduction to Ottawa

Ottawa is nested inside a statistical jurisdiction known as Ottawa-Gatineau, which is the only major metropolitan area in Canada to bridge two provinces. Ottawa-Gatineau is a bilingual city, with many commuters traveling between the French-speaking province of Québec and the English-speaking province of Ontario daily to live and work. Simultaneously, the federal planning agency known as the National Capital Commission has jurisdiction for the National Capital Region, which roughly approximates the territory of the Ottawa-Gatineau Census Metropolitan Area or CMA (as defined by Statistics Canada).[1]

Figure 1. The City of Ottawa *vis-à-vis* Ottawa-Gatineau

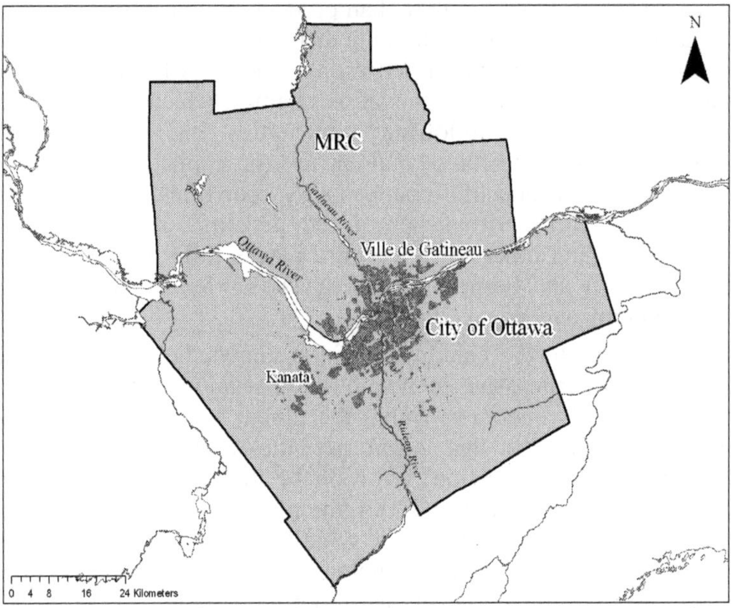

The Ottawa - Gatineau CMA in 2006

Source: John Tinholt, 2007.

Ottawa-Gatineau is Canada's fourth largest CMA after Toronto, Montréal and Vancouver. Using the 2001 Census boundaries, the population of Ottawa-Gatineau grew steadily between 2001 and 2005. Meanwhile, as a destination for immigrants to Canada, Ottawa-Gatineau does not rank as high as Toronto or Vancouver, which explains why its growth rate is lower than other major cities. However, the type of immigrants that do migrate to Ottawa-Gatineau are more likely to have degrees in the physical sciences, engineering and trades than those moving to other major Canadian cities (Strategic Research and Statistics, 2005).

1 The authors gratefully acknowledge the map generated by John Tinholt for this paper (using Geogratis data). The gray-toned area is the census metropolitan area, with the City of Ottawa to the south of the Ottawa River, and the Ville de Gatineau to the north.

Table 1. Population Growth in Canada's Four Largest Cities

Year	Ottawa-Gatineau	Toronto	Montréal	Vancouver
2001	1,102,900	4,883,800	3,507,400	2,076,100
2002	1,118,800	5,020,400	3,547,100	2,111,300
2003	1,131,800	5,116,700	3,578,800	2,142,100
2004	1,141,400	5,214,000	3,609,600	2,174,000
2005	1,148,800	5,304,100	3,635,700	2,208,300
Growth: 2001-05	4.2%	8.6%	3.7%	6.3%

Source: Statistics Canada (2006a), CANSIM Table 051-0034.

Median incomes in Ottawa-Gatineau are also higher than in other major Canadian cities.

Table 2. Median Incomes for Canada's Major Cities (in Cdn$)

	Median Incomes for 2001 (in Cdn$)
Toronto	25,593
Montréal	21,888
Vancouver	23,237
Ottawa-Gatineau	28,956
City of Ottawa	29,978

Source: Statistics Canada (2006b).

Overall, Ottawa has certain built-in advantages as a high-technology centre (e.g., its educational achievement levels, its quality of life). And, some of these advantages extend beyond the purview of advanced technology. For example, KPMG has done an international survey of business competitiveness across all G-7 nations called the Competitive Alternatives Study: 2006 Edition (KPMG, 2006). The industries studied in the 2006 edition include the following: aerospace, agri-foods, automotive, chemicals, electronics, medical devices, metal components, pharmaceuticals, plastics, precision manufacturing, telecommunications, biotechnology, clinical trials, product testing, software design, and Web, multimedia and corporate service centres. These high-tech centred industries were tested for costs related to labour, taxes, real estate, and utilities. Canada ranked first among G-7 nations in terms of cost-competitiveness. Ottawa performed ahead of Toronto and Vancouver, but not as well as Montréal due to lower real estate costs in that city.

Because Ottawa is Canada's capital, its economy is unique among other Canadian cities. Federal employment has grown by nearly 10% since 2001. It is worth noting that government employment is atypical in that it is simultaneously stable over time (i.e., resistant to the business cycle), well-paying, and knowledge-intensive. However, one exception in Ottawa's past is Programme Review, which involved the contraction of the federal civil service and hit Ottawa particularly hard through the loss of nearly 25,000 jobs over 5 years starting in 1993. However, since the late-1990s, federal government employment has been on an upswing.

Table 3. Federal Employment in Ottawa-Gatineau: 2001-2005

Year	Federal Government Employment in the Ottawa-Gatineau CMA
2001	103,685
2002	108,050
2003	112,310
2004	112,324
2005	113,835

Source: Statistics Canada (2006c). CANSIM Table 183-0003.

Of total federal employment in Canada, 41.1% of the jobs were in Ottawa-Gatineau in 2005 (Statistics Canada, CANSIM Table 183-00003, 2006). Of the entire federal budget for wages and salaries in 2005 of 1,281,990,000 (Cdn$), 45.6% of those wages and salaries were paid in Ottawa-Gatineau, indicating that the wage proportion outstrips the employment proportion by approximately 4.5%. This supports the presupposition that higher-end federal jobs would be located in Ottawa-Gatineau. Meanwhile, these figures do not include the substantial number of people that are hired on contract or temporarily by the federal government.

Along with the employment engine of the federal government, advanced-technology employment in Ottawa-Gatineau, and particularly in the City of Ottawa, has been competitive. At the peak of the advanced-technology boom in 2001, high-tech employment reached 79,000 (OCRI, 2005) and for the first time approximately matched federal employment on the Ottawa side of the Ottawa-Gatineau CMA (Harrison, Cooper and Mason, 2004). However, approximately 10,000 technology-related jobs were lost following the stock market crash and high-tech sector downturn over the summer of 2001, leaving only 69,500 high-tech employees in 2003 (OCRI, 2005). Once again, the federal government became the City of Ottawa's dominant employer. Currently, high-tech employment has not quite recovered to levels experienced in 2001, but it remains the area's major non-governmental employment sector. In 2005, the federal government remains the key driver of Ottawa-Gatineau's economy with over 113,000 civil servants, but now operates in tandem with 76,126 advanced-technology industry employees (OCRI, 2005). It is worth noting that these employment figures do not reflect federal employees employed in advanced technology and research.

Though Ottawa is sometimes pejoratively referred to as *a government town*, the federal government is a rich source of both employment and expertise. Its research institutes are centres of excellence that benefit Ottawa businesses directly. As both a buyer of technological products and services and a collaborator in science and technology, the Government of Canada is invaluable to Ottawa's economy. It is an economic driver in the strongest sense – greater than any other growth cluster. Far from having a limiting effect on Ottawa's economy, the federal government's presence (and especially its research centres) represents the genesis of Ottawa's technology boom. From those beginnings, technology giants grew and spawned countless other companies and capabilities. But how did Canada's capital city become *Silicon Valley North*? How was the local advanced-technology industry hatched?

The History of Ottawa's Advanced-Technology Sector[2]

The region's powerful research strengths have twin roots. The first lies in the growth after World War II of public scientific research labs, including the National Research Council (NRC), the Defence Research Establishment, the Communications Research Centre (CRC) and later, the Atomic Energy of Canada Laboratory (AECL). The second root lies in a decision taken in the late 1950s by Northern Electric (now Nortel) to choose Ottawa as the site for its small but rapidly growing R&D operation.

During the war, the National Research Council (NRC) built the foundation for its development into a world-class research institution, developing expertise in fields like chemistry, communications, engineering, information systems and later biotechnology. The origins of more than 60 technology-based companies can be traced back to the NRC, beginning in 1950 with Computing Devices of Canada (known as CDC and now General Dynamics Canada). CDC started with the development of a navigational aid for aircraft and grew into a major technology player making a wide range of highly sophisticated military products. As viewed by advanced-technology pioneers, if Ottawa had not been the location of government laboratories such as the National Research Council and others at the end of World War II, firms like Computing Devices of Canada and Leigh Instruments would not have been created. By attracting talent and developing a range of new technologies, these entities in turn supported the start up of at least another dozen firms – Dy 4 and Lumonics among them.

---

2 This section is based on Taylor (1986), Keshen and St. Onge (2001), and Holzman (1994).

Through the 1960s, 1970s and 1980s, the federal government continued to influence the development of the Ottawa advanced-technology cluster. By the mid-1990s, federal R&D expenditures had risen to 5 billion Cdn$ annually and about 30% of that amount was concentrated on activities in Ottawa. In strategic collaboration with local companies, the NRC's institutes and the CRC's labs supported innovative developments in communications, microelectronics, software and, later, photonics technologies, and provided incubator facilities to sponsor early stage growth. In the 1960s and 1970s, the federal government also used its mainframe computer purchasing requirements to encourage multinational corporations to establish R&D and manufacturing facilities in Canada. Digital's establishment in Ottawa, which eventually produced a number of spin-offs, and the IBM origins of Ottawa-born SHL Systemhouse, are two examples of the kind of development that was stimulated by the investment of foreign corporations during that early pre-NAFTA (North American Free Trade Agreement) period.

Linkages also developed among public research institutes, the educational sector and local business. As they expanded through the 1970s and 1980s, universities and colleges in Ottawa became important R&D players. Ottawa also attracted a provincial Centre of Excellence, Communications and Information Technology Ontario (CITO), as well as federally-funded, industry-led consortia like CANARIE. All of this activity furnished the region with a public sector R&D infrastructure that was unparalleled in the country.

The greatest single influence on the development of the Ottawa cluster, however, was the decision by Northern Electric (Nortel) to concentrate its R&D activity in the nation's capital via two subsidiaries – Bell Northern Research (BNR) and Microsystems International. BNR's labs and test facilities seeded the development of Ottawa's telecom expertise with research focused on the use of computers in electronic switching. And, Microsystems was an early developer of semi-conductor technology. Although it was shut down in the mid-1970s, Microsystems attracted many highly-skilled and entrepreneurial engineers and scientists to Ottawa. Some of these later joined BNR, helping to spur Nortel's research operation into one of the most powerful and innovative R&D centres in the world – one which spent and spends billions of dollars, employs thousands of highly-skilled people, and spun off several dozens of companies, including the future fibre-optics powerhouse JDS.

The Ottawa story emphasises the critical importance in cluster development of a deeply rooted R&D strength. It also clearly underscores the fact that access to technology demands the presence of world-class scientific research institutions. Only through the combined impact of local public and private sector institutional research activity could Ottawa have spawned its own homegrown high-tech industry.

Capitalising on the talent and funding available in the region, the 1970s start-ups like Mitel, Gandalf, SHL Systemhouse and Quasar (now Cognos) grew steadily into the early 1980s, when they were joined by other Nortel – or CDC – or Leigh-inspired spin-offs. These companies and others developed a strong track record as a result of working with local institutions to commercialise new discoveries and of converting their R&D investments into products and services for a global market. It took decades, but by the mid-1980s, when the controlling interest of Mitel had been sold to British Telecom and Mitel's co-founders had moved on to start Corel and Newbridge, the Ottawa cluster began to take off.

By the early 1990s, the companies in the region had achieved critical mass. This phenomenon and the business opportunities it presented acted in turn as a magnet for large foreign corporations like Alcatel, Cadence, Cisco, Nokia and Siemens. These firms and other world players moved in with major R&D investments, aggressively seeking the partnerships, skilled personnel and the innovative environment they needed to support their own corporate expansions. By 2001, before the dot.com market correction, the region was home to nearly 1,000 technology companies and employing approximately 79,000 people (OCRI, 2005).

Ottawa Today

Compared to 2000, Ottawa in 2006 has more companies but fewer tech workers. Approximately 76,000 people work in one of the 1811 high-tech industries in the Ottawa-Gatineau area (OCRI, 2005). Today, Ottawa's high-tech cluster competes primarily with cities in the United States. Like other cities in Canada, much of Ottawa's trade is directed south rather than east-west. The cities that Ottawa competes with specialise in the same export-oriented industries as Ottawa does. Its key competitors are mainly Raleigh-Durham in North Carolina, San Francisco in California, and Austin in Texas. While providing competition, Ottawa's sub-clusters also collaborate with these cities where synergies and opportunities exist (Tremblay, 2002).

Ottawa's key high-tech sub-clusters are technology services; software; telecommunications; defence and security; microelectronics/ wireless; and photonics (OCRI, 2005). The software and communications sub-clusters provide an array of packaged software and information technology (IT) services, as well as certain telecommunication services to domestic and international markets. Many of Ottawa's software firms are among the largest in Canada, and hundreds of smaller and younger firms contribute to the growth of the cluster. More specifically, the telecommunications sub-cluster has been at the forefront of the technology sectors driving Ottawa's economy. It covers a wide range of technical activities from the assembly of basic telephone hand sets to highly-sophisticated digital switches, with more than 11% of the city's high-tech firms and more than 15% of local high-tech employment (OCRI, 2005). 90% of all of Canada's industrial telecommunications R&D takes place in Ottawa, facilitated by the presence of the National Research Council's Institute for Information Technology and the Communications Research Centre.

Photonics, meanwhile, is a high-profile growth area. In 2005, it represented 6% of the area's high-tech firms and over 10% of high-tech employment. Regarding photonics, in 2000, there were 75 firms in Ottawa employing over 13,500 people. Ontario is home to about 80% of Canada's photonics companies and over half of those companies are in Ottawa. As optical technology has evolved quickly along with growing investment, the Ottawa photonics sub-cluster was created to advance the industry. Technological synergy between photonics, tele-communications and microelectronics is strong and is one key reason why Ottawa leads the country in the growth of these sub-clusters.

The life sciences sub-cluster, which covers such fields as biotechnology, medical technology and health-related systems, employs more than 13% of the high-tech employees in the region (OCRI, 2005). Spawned by federal government organisations and local institutions such as the National Research Council, the Ottawa Health Research Institute and the University of Ottawa Heart Institute, the private sector has expanded to more than 100 life sciences and allied companies. The 22-acre Ottawa Life Sciences Technology Park adjoining the Ottawa Health Sciences Centre was established by the Province, the City, and local hospitals. The Ottawa Biotechnology Incubation Centre was established and approximately 20 related research institutes exist in the region, investing approximately 350 million Cdn$ in R&D each year. The Ottawa Life Sciences Council, a non-profit corporation representing the sector, provides leadership in this cluster. In combination, these myrid sectors represent the private component of Ottawa's New Economy.

Ottawa's New Economy

Using a framework provided by the Progressive Policy Institute (Atkinson and Gottlieb, 2001), the measurement of whether the New

Economy is present in a city depends on the presence of five central characteristics: knowledge jobs, globalisation, economic dynamism, the digital economy and innovation capacity. *Knowledge jobs* can be defined as managerial, professional and technical positions held by individuals with some university training. Across the globe, required skill levels for employability are increasing in all sectors. Three measurements for knowledge jobs are: office jobs (the percentage of the workforce working in offices as opposed to on the farm, in industry or in retail); managerial, professional, and technical jobs; and levels of educational achievement of the labour force. As demonstrated in the data for each of these three variables, the City of Ottawa demonstrates the highest presence of knowledge jobs among Canada's largest cities.

Table 4. Office Jobs in Canada's Largest Cities in 2001

Office Jobs	Toronto	Montréal	Vancouver	Ottawa-Gatineau	City of Ottawa
Finance and Real Estate	9.2	6.1	7.5	4.9	5.2
Health and Education	13.2	16.5	16.6	15.8	15.4
Business Services	23.7	21.1	23.7	22.1	24.2
Other Services	15.3	16.8	19.2	31.4	30.1
Total Proportion	61.4%	60.5%	67.0%	74.2%	74.9%

Source: Statistics Canada (2005).

Management, professional and technical jobs, insofar as they are reflected by Statistics Canada's categories of (1) Management, (2) Natural and Applied Sciences, and (3) Social Science, Education and Government, are better represented in Ottawa than in Canada's other largest cities as well.

Table 5. Management, Professional and Technical Jobs in Canada's Largest Cities

Management, Professional & Technical	Toronto	Montréal	Vancouver	Ottawa-Gatineau	City of Ottawa
Management	13.0	10.8	11.9	12.9	13.8
Natural & appl'd science	8.2	7.4	7.1	12.9	14.7
Social science, education & government	7.5	8.2	8.2	10.6	10.7
Total Proportion	28.7%	26.4%	27.2%	36.4%	39.2%

Source: Statistics Canada (2005).

In terms of levels of educational achievement, the City of Ottawa also outperforms Canada's other largest cities by a margin of 4% when compared to Toronto, the next best performer. This is a function of the presence of the federal government, and the presence of a number of good universities and job prospects.

Table 6. Educational Achievement in Canada's Largest Cities

% of Population with a University Education	Toronto	Montréal	Vancouver	Ottawa-Gatineau	City of Ottawa
20-34 cohort	32.7	27.4	29.5	33.4	37.0
35-44 cohort	32.5	27.2	29.7	33.8	37.9
45-64 cohort	27.2	23.1	27.0	31.8	35.6

Source: Statistics Canada (2006d).

As shown, Ottawa-Gatineau performs better than the other large Canadian cities in terms of knowledge jobs. Meanwhile, *globalisation* involves actions or processes that involve the entire world and result in making something worldwide in scope. Globalisation pulls international groups of people into greater economic and cultural interaction with each other. Globalisation is the one weakness in Ottawa-Gatineau's armour. For example, foreign direct investment (FDI) is falling. The Conference Board of Canada (2004) has attempted to ascertain what is going on, and found that the three factors that cause the most problems for Canada's rate of FDI involve the following: local suppliers, the state of the business infrastructure, and the tax and regulatory environment. This is the viewpoint of 100 executives at multinatinal corporations and needs to be considered within the confines of their perspective: profit maximisation. That said, FDI is not necessarily critical if funds are being generated within the country and venture capital can still be raised.

Economic dynamism relates to economic competition and involves the fast-growing, entrepreneurial companies that are using innovation and the ability to get to the market faster as a determinant of competitive advantage. The new *gazelle firms* (firms that have had sales growth of 20% or more for four straight years) are correlated with growth in overall employment. Job churning, which is regarded as one dimension to economic dynamism and is the product of both new business start-ups and business failures, has become apparent in Ottawa. This can be demonstrated by the high number of business starts as well as bankruptcies. Regarding bankruptcies up to the dot.com crash, Ottawa-Gatineau consistently had more than the other largest Canadian cities. Bankruptcies in this context are used as a proxy for risk taking and entrepreneurial spirit. When combined with healthy numbers of business start-ups, they are a good indication of job churning – which suggests economic dynamism. Between 1996 and 2001, new business registrations in Ottawa-Gatineau remained consistent between a low of 5,698 in 1996 and the high in 1998 of 7,053.

Table 7. Commercial Bankruptcies per 1,000 Companies by Metropolitan Area Plus the Canadian Average

	1996	1997	1998	1999	2000	2001
Toronto	11	10	9	7	7	9
Montréal	20	17	14	13	11	11
Vancouver	6	5	5	5	5	6
Ottawa	20	20	16	15	12	12
Canada	15	13	11	10	10	10

Source: Centre on Governance (2002).

The *digital economy* is another facet of the New Economy and refers to individual, business and government transactions (e-commerce, information) that are conducted through digital electronic means. Measures for the digital economy include the on-line population, technology in the schools, and the presence of digital government. Regarding the on-line population, Ottawa has the highest proportion of Internet users by household (65.2%) of all of Canada's major urban centres. Toronto had a proportion of 57.9% and Montréal had a proportion of 46.6% (City of Ottawa, 2003). More recent research by Statistics Canada confirms this (OCRI, 2005).

Regarding the use of technology in schools, e-learning is already a major force at Canadian universities, colleges and K-12. Even in 2000, over 55% of Canada's colleges and universities were offering distance education (e-learning) as part of their curricula. Virtually 100% of schools in Ottawa-Gatineau are linked to high speed internet through fibre optic cable (OCRI, 2005). The use of the internet in schools opens the door to demands for digital government (e.g., the use of digital technologies to deliver government services). Within the City of Ottawa's Broadband Strategy (City of Ottawa, 2003b), the provision for e-democracy, e-government and e-services are addressed. One of the goals of the 20/20 Strategy is the aspiration that "live, two-way video will allow users to see and discuss issues with City Councilors without leaving their neighbourhood." The advent of e-government in Ottawa was the web-cast of the City's 20/20 Strategy in 2002. Nearly all of the city's current plans, news and records of council meetings are already on-line and readily available to civil consumers.

Finally, *innovation capacity* refers to the ability of a city to generate new ideas, new knowledge and new technologies, as well as to adopt them. Innovation capacity can be measured by looking at technology-related jobs (i.e., the share of jobs in advanced technology industries and services); the presence of scientists and engineers; patent generation; investment in Research and Development; and the availability of venture capital.

The Ottawa Centre for Research and Innovation (OCRI) maintains data on high-tech employment in Ottawa. Since the employment trough in 2001, high-tech employment has been steadily improving over time. The January, 2006, levels are nearly back to where they were at the peak of the boom in 2001. High-tech employment in this case is disaggregated from the federal and provincial governments and reflects only those high-technology workers operating within private industry.

Table 8. Number of High-tech Employees in Ottawa-Gatineau: 1996-2006

Year	Number of High-tech Employees
1996	40,000
1997	48,312
1998	51,000
1999	52,000
2000	65,000
2001	79,000
2002	69,500
2003	66,500
2004	63,700
2005	67,800
2006	76,126

Source: OCRI (2005).

Despite the sector downturn in 2001, growth in the number of high-tech companies in Ottawa has steadily increased since 1996. In 2006, over 50 companies in Ottawa have over 500 employees, which is a 200% increase over the 2000 benchmark when only 17 companies were that size.

Table 9. Number of high-tech companies in Ottawa-Gatineau: 1996-2006

Year	Number of High-tech Companies
1996	700
1997	815
1998	900
1999	903
2000	980
2001	1,000
2002	1,100
2003	1,430
2004	1,500
2005	1,688
2006	1,811

Source: OCRI (2005).

The largest sub-clusters in Ottawa can be ranked in the following way according to the number of companies and their percentage of total private high-tech employment:

1. Technology services: 37% of companies (management information services, system integrators, consultants and administrative services) and 22% of high-tech employment.
2. Software: 20% of companies and 15% of high-tech employment.
3. Telecommunications: 11% of companies and 15% of high-tech employment.
4. Defence and security: 8% of companies and 10% of high-tech employment.
5. Microelectronics/wireless: 7% of companies and 11% of high-tech employment.
6. Photonics: 6% of companies and 10% of high-tech employment (OCRI, 2005).

The presence of scientists and engineers within the workforce is also important because they can act as innovation drivers seeking new solutions to existing or emerging problems. For immigrants with college or university educations living in the Ottawa-Gatineau CMA that came to Canada after 1996, the proportion of male immigrants having degrees in the physical sciences, engineering and trades is 72% as compared to Canadian-born men at 42% (Strategic Research and Statistics, 2005). University and college-educated women that are immigrants to Ottawa-Gatineau are also more likely to have degrees in the physical sciences and engineering (at 38% versus 13% for Canadian-born women). This stresses the importance of Canada's point system in not only making Canadian cities more cosmopolitan, but also filling gaps in the Canadian graduation pattern. Immigrants living in Ottawa-Gatineau are much more likely to be educated in the sciences than immigrant cohorts living in other major Canadian cities. More generally, in 2001, in the category of Professional, Scientific and Technical Services, Ottawa had 5.6% of its population employed, as compared to 5.0% for Toronto, 4.0% for Montréal, and 4.4% for Vancouver (Statistics Canada, 2001).

With higher levels of scientists and engineers, it is not surprising that patents per capita levels are higher in Ottawa-Gatineau than elsewhere in the country. Patents per capita data are not available by CMA from Statistics Canada or the Canadian Intellectual Property Office (CIPO). However, nearly 70% of Canadian manufacturing firms apply for a patent in both Canada and the US, and the US does maintain patent information by metropolitan area. A methodology proposed by the Centre for Governance at the University of Ottawa suggests that Canadian patents per capita can be estimated by taking the US data and multiplying it by 1.3. Using this methodology, patents per capita are highest in Ottawa-Gatineau, followed by Vancouver, Montréal and Toronto. Patents per capita levels in Ottawa-Gatineau are more than twice those generated in either Montréal or Toronto (Centre for Governance, 2002).

To help generate innovation and subsequent patents, there needs to be investment in research and development along with available public/private or venture capital. The research and development picture in Ottawa-Gatineau is strong. An Ottawa company, Nortel Networks, single-handedly spent 2.5 billion Cdn$ on telecommunications research (Research Infosource, 2006). Nortel is joined by other companies heavily involved in R&D like Alacatel, Cognos, Zarlink, Mitel and Tundra Semiconductor. The federal government also invests in R&D, and invested 60 million Cdn$ to help Mitel Networks and March Networks to research broadband communications technologies (City of Ottawa, 2003b).

Venture capital supports innovation by making funding available for research and entrepreneurship in products and services that have higher-than-average levels of risk. The success of Silicon Valley is well-documented and is far more related to the banking and finance industries making venture capital more readily available in that area rather than anything specifically related to the San Francisco Bay Area like its university nexus. Venture capital in Ottawa-Gatineau has been variable over the past decade. But, the most recent trough seems to be over and 2005 experienced the first upswing in venture capital investment since the downturn in 2001. For 2005, Ottawa received 43% of all venture capital invested in Ontario and 28% of the Canadian total (OCRI, 2005).

Table 10. Venture Capital Amounts for Ottawa: 1996-2005

Year	Total Venture Capital in 000,000s Cdn$
1996	60
1997	190
1998	75
1999	275
2000	1,350
2001	1,100
2002	735
2003	287
2004	248
2005	362

Source: OCRI (2005).

Since the creation of a constellation of factors for innovation is the launch pad for the knowledge city, the amount of available venture capital is particularly important. An exploration of innovation engines is warranted, but is outside the confines of the current study. Certainly, investigation into innovation engines like the following variables is warranted: universities, big events, libraries, museums, digital infrastructure, post-industrial science parks, brownfield industrial zones, urban virtual space, and even coffee shops or cafés and other so-called *third places* (Dvir and Pasher, 2004).

Conclusions

Building a technology cluster is the work of decades. To the outside world, Ottawa looks like an overnight success story. But the groundwork had actually been laid 30, 40 and even 50 years earlier. This Ottawa case study demonstrates that its R&D strength and related patent generation are the most important factors in the formation of the cluster, not only because it spurs a collective capacity to generate innovative ideas for commercial applications, but also because creative and talented people go where there is technological challenge and excitement. Due to local world-class research institutions like the NRC, and good educational institutions like the University of Ottawa and Carleton University, which provided access to technology while attracting highly-qualified and entrepreneurial people, Ottawa was able to consolidate the essential elements of making a globally-competitive knowledge city.

With its R&D base, pool of skilled labour and, especially, the vision, drive and persistence of its local entrepreneurs, Ottawa was also able to transcend its one real disadvantage – the lack of availability of early stage financing. Access to venture capital never played a significant role in the formation of the Ottawa sub-clusters, which suggests that this element may be a less critical factor in cluster formation when the other elements – in particular, the entrepreneurial element – are there. Ottawa's visionary enterprises created new models to finance and mentor start-ups, while also stimulating and supporting the development of a powerful community technology culture. This culture – driven by the expansion of linkages among business, governmental and educational players – helped to provide the infrastructure needed for success in innovation.

References

Atkinson, R., Gottlieb, P. (2001). The metropolitan new economy index: benchmarking economic transformation in the nation's metropolitan areas. *The New Economy Index*. Retrieved May 14, 2006 from http://www.neweconomyindex.org/ metro/index.html.

Batten, D.F. (1995). Network cities: creative urban agglomerations for the 21st century. *Urban Studies, 32* (2), 313-327.

Centre on Governance. (2002). *Ottawa works: a mosaic of Ottawa's economic and workforce landscape. Report 1: Ottawa's workforce environment*. Ottawa: Human Resource Development Canada.

City of Ottawa. (2003a). Map of the national capital region. Retrieved May 22, 2006, from http://ottawa.ca/cityservices/planningzoning/2020/window/ maps _en.shtml.

City of Ottawa. (2003b). *Ottawa 20/20 – broadband plan. Final draft*. Ottawa: City of Ottawa.

Conference Board of Canada. (2004). *Canada is losing its advantages as a destination for foreign direct investment*. Ottawa: Conference Board of Canada.

De la Mothe, J., Paquet, G. (1999) *Information, innovation, and impacts*. Boston: Kluwer.

Dvir, R., Pasher, E. (2004). Innovation engines for knowledge cities: an innovation ecology perspective. *Journal of Knowledge Management, 8* (5), 16-27.

Edvinsson, L., Dvir, R., Pasher, E. (2004). Innovations: the new unit of analysis in the knowledge era-the quest and context for innovation efficiency and management of IC. *Journal of Intellectual Capital, 5* (1), 40-58.

Egazakis, K., Metaxiotis, K., Psarras, J. (2004). Towards knowledge cities: conceptual analysis and success stories. *Journal of Knowledge Management, 8* (5), 5-15.

El-Diraby, T., Wang, B. (2005). E-Society Portal: Integrating Urban High Construction Projects into the Knowledge City. *Journal of Construction Engineering and Management, 131* (11), 1196-1211.

Harrison, R., Cooper, S., Mason, C. (2004). Entrepreneurial activity and the dynamics of technology-based cluster development: the case of Ottawa. *Urban Studies, 41* (5/6), 1045-1070.

Holzman, J. (1994). *An economic profile: Ottawa*. Ottawa: Office of the Mayor, City of Ottawa.

ICF Consulting. (2000). *Choosing a future: a new economic vision for Ottawa: final report*. Ottawa: Ontario Ministry of Economic Development and Trade, Office for Urban Economic Development.

Keshen, J., St. Onge, N. (2001). *Ottawa: making a capital*. Ottawa: University of Ottawa Press.

Komninos, N. (2002). *Intelligent cities: innovation, knowledge systems and digital spaces*. London: Spon Press.

KPMG. (2006). *Competitive alternatives: KPMG's guide to international business costs*. 2006. Edition. Retrieved May 6, 2006 from http://www.mmk consulting.com/compalts/reports/2006_compalt_report_vol1_en.pdf.

Molitor, G. (2000). Five economic activities likely to dominate the new millennium: I. Introduction. *Technological Forecasting and Social Change*, *64*, 271-276.

Molitor G. (2005). Five economic activities likely to dominate the new millennium. VII: Principles and patterns of economic era development. *Technological Forecasting and Social Change*, *72*, 85-99.

OCRI (Ottawa Centre for Research and Innovation). (2003). *Ottawa 2003: technology industry guide*. Ottawa: OCRI.

OCRI (Ottawa Centre for Research and Innovation). (2005). *The OCRI report: 2005. Economic, technology and educational indicators*. Ottawa: OCRI.

Picon, A. (1998) *La ville territoire des cyborgs*. Paris: Les Éditions de L'Imprimeur.

Polèse, M., Tremblay, R. (2005). L'économie du savoir et la manie des "rankings": une analyse intégrée des villes canadiennes et américaines. *Le géographe canadien/Canadian Geographer*, *49* (2), 198-213.

Research Infosource. (2006). *Canada's corporate innovations leader*. Research Infosource. Downloaded on May 10, 2006 from http://www.researchinfosource.com/2005-top100-sup.pdf.

Statistics Canada. (2001). *Ottawa-Hull community profile*. Statistics Canada. Retrieved on May 14, 2006 from www.statcan.ca.

Statistics Canada. (2005). Experienced labour force 15 years and older by industry, by census metropolitan areas (2001 Census). Statistics Canada. Retrieved on May 14, 2006 from http://www40.statcan.ca/l01/cst01/labor47e.htm?sdi=experienced% 20labour%20industry.

Statistics Canada. (2006a). *Population of census metropolitan areas* (2001 census boundaries). Statistics Canada. Catalogue Numbers 91-213-XIB and 91-213-XPB plus CANSIM Table 051-0034. Ottawa: Statistics Canada.

Statistics Canada. (2006b). *Community profiles*. All data for each CMA. Income in 2000, median total income of persons 15 years of age and over. Ottawa: Statistics Canada.

Statistics Canada. (2006c). *Federal employment in Ottawa-Gatineau: 2001-2005*. CANSIM Table 183-0003. Ottawa: Statistics Canada.

Statistics Canada. (2006d). *Detailed highest level of schooling (20), age Groups (13B) and sex (3) for population 15 years and over, for Canada, provinces, territories, census metropolitan areas and census agglomerations*. 2001 Census – 20% Sample Data – Catalogue No.97F0017XCB2001001. Ottawa: Statistics Canada.

Strategic Research and Statistics. (2005). *Recent immigrants in metropolitan areas*. Catalogue No.MP22-20E/12-2005E-PDF. Ottawa. Citizenship and Immigration Canada.

Taylor, J. (1986). *Ottawa: an illustrated history*. Toronto: James Lorimer.

Tremblay, R. (2002). *La passion pour les classements: revue des principaux « rankings » des technopoles américaines et canadiennes et américaines.* Montréal: INRS.

Tremblay, R. (2003). *Le tandem Gatineau-Ottawa ou la frontière fantôme*. Dans Coté, R. et Venne, M. (dir.). Montréal: Fides, 240-248.

Tremblay, R. (2004a). Revue des principaux rankings des technopoles nord-américaines. *Revue canadienne des sciences régionales/Canadian Journal of Regional Sciences*, *27* (1), 121-134.

Tremblay, R. (2004b). La place d'Ottawa-Gatineau parmi les villes du savoir nord-américaines. *Organisations et territories*, *12* (3), 12-20.

Van den Besselaar, P., Melis, I., & Beckers, D. (2000). Digital cities: organization, content and use. In T. Ishida & K. Isbister (eds.). *Digital cities, experiences, trends, and perspectives*. Lecture notes in computer science 1765. Berlin: Springer, 18-32.

Work Foundation. (2006). The work foundation identifies best UK cities in which to live, work and invest. *Work Foundation*. Retrieved May 20, 2006. http://www.theworkfoundation.com/newsroom/pressreleases.jsp?ref=190.

CHAPTER 2

Ottawa – North American Technopole

Fact or Fiction?[1]

Rémy TREMBLAY

Canada Research Chair on Knowledge Cities,
Université du Québec à Montréal

Introduction

Municipalities, business associations and several economic development groups around Ottawa[2] have the feeling that their region belongs with the top ranking high-technology centres in North America[3]. Indeed many high profile companies have their headquarters, or a regional office, or even research and development (R&D) facilities in the Ottawa region: Corel, Cognos, IDS, Uniphase, Adobe, Nortel, Mitel and Cisco Systems. In addition, statistics tend to support, at least in part, the beliefs or statements of business developers (Tremblay, 2004). However, does this perception rendered by those who work in the field actually promote Ottawa to the status of a prominent technopole on the continent?

The objective of this chapter is to reveal the ranking of Ottawa, Canada's capital, among the 24 other large North American high-tech cities,

---

1 This study was carried out with the support of the Canada Research Chair in Urban and Regional Studies headed by professor Mario Polèse, Université du Québec, Institut national de recherche scientifique (INRS).

2 For the purpose of this study, Ottawa includes the census metropolitan area (CMA) of Ottawa-Gatineau with its population of 1,146,000 in 2004 (Statistics Canada, 2005).

3 *High technology* as a concept lacks clarity (Tremblay & Rolland, 2003; Fache, 2002; Mille & Adams, 2001). The definition employed is based on the importance of the jobs created in the professional, scientific and technical services in given communities (Appendix C).

making use of statistics compiled in 2004 and referring to a survey on the perception of the actors themselves concerning the significance of Ottawa in this respect. First, the demographic, social, cultural and economic metamorphosis experienced in and around Ottawa is outlined. Second, the results of a web survey – a polling of over 1,500 local high-tech entities – are presented. Finally, the survey results are compared to specific data generated in another, previous study (Polèse and Tremblay, 2005).

Ottawa: A Transformed City

Over the past decade, Ottawa's image has evolved from a drab and even stern city to that of a conurbation displaying major demographic, social, cultural and economic diversification (Tremblay, 2003). In 2001, 1,108,500 people lived in the Ottawa area, up 1.9% over the preceding year; only three Canadian metropolitan areas were recording a higher demographic rate of growth: Toronto, Calgary and Windsor. Again in 2001, 17.6% of the Ottawa population was born outside Canada (21.1% in Ottawa and 6.6% in Gatineau), compared to 43.7% in Toronto, 18.4% in Montréal and 2.9% in Québec City. From coast to coast, Canada has an average of 18.4% of its residents being foreign-born.

Culturally and socially, Ottawa is now on par with other North American cities of equivalent size (ICF Counsulting, 2000; de la Mothe and Paquet, 1999). Ottawa residents and local media now consider Ottawa's lifestyle to be one of young IT professionals and engineers, rather than that of civil servants. Luxurious condominiums and lofts are in high demand despite their price tag (300,000 to 2 million Cdn$), along with trendy cocktail lounges and *boutique hotels*. In the summer of 2005, no fewer than 20 such community sites were under construction.

Although the public service sector remains prominent, the economy of the national capital has undergone major changes (Holzman, 1994; Shavinia, 2004). Indeed, since 1990, the high-tech sector has grown at such a pace that in 2000, according to Mallet (2002), 80,000 individuals were on the payroll of knowledge-economy businesses, almost as many as in government offices. Between 2000 and 2004, the meltdown of the high-tech sector caused harm to Ottawa's economy. In 2001, nearly 61,000 engineers, technicians and other specialists were working in the 1,500 high-tech firms which is, given the size of Ottawa, a substantial component of its workforce (Statistics Canada, 2001; Centre on Governance, 2003). Polèse and Tremblay (2005) confirm and demonstrate that Ottawa ranks among 20 cities[4] enumerating a population of 500,000 or more in which a significant proportion of the workforce is active in professional, scientific and technical fields (Table 1[5]). Despite the sector downturn in 2001, Ottawa still ranks in the top ten North American cities where a high percentage of people holding university (bachelor and PhD) degrees.

Table 1. Percentage (%) of the Workforce in Professional, Scientific and Technical Services in the United States and in Canada (top 20)

Rank	Cities	% of workforce
1	Washington-Baltimore	11.54
2	San Francisco	10.95
3	Calgary	10.91
4	*Ottawa*	*10.47*
5	Toronto	9.78
6	Raleigh	9.74
7	Denver	9.14
8	Boston	8.98
9	Albuquerque	8.76
10	Vancouver	8.74
11	San Diego	8.72
12	Austin	8.55
13	New York	8.20
14	Atlanta	8.19
15	Montréal	7.95
16	Colorado Springs	7.92
17	Minneapolis	7.74
18	Chicago	7.70
19	Houston	7.44
20	Philadelphia	7.40

Source: Statistics Canada (2001) and US Census (2000).

In short, the materialisation of employment opportunities that require specialised technical knowledge has introduced a shift in Ottawa's economic momentum (Shavinina, 2005). Nevertheless, what is the perception of these highly skilled workers regarding Ottawa as 'Silicon Valley North'? Does the mental representation held actually match the available statistics, or does the perception of those concerned nourish another brand of reality? And, in the minds of the actors, is Ottawa a member of the select club of North American high-tech communities? The perception of high-tech business people in Ottawa is herein com-

4 In this article, the definition of Canadian and American cities includes its surrounding suburbs. Canadian terminology involves CMAs (Census Metropolitan Areas), which are analogous to CSAs (Consolidated Statistical Areas) in the United States.

5 See Appendix C for references and explanations relating to the tables.

pared with Statistics Canada and US Census data depicting the same set of cities.

Methodology of the Study

In order to identify and contact the companies located in the Ottawa area, permission to use the OCRI database (Ottawa Centre for Research and Innovation) was secured. OCRI is a non-profit, partnership organisation programmed to enhance Ottawa's globally competitive knowledge economy with a view to promote, attract and support high-tech businesses with a stake in the Ottawa region. OCRI brings together more than 1,500 stakeholders and members. The Ottawa Centre organises seminars and training sessions; it publishes facts and surveys focusing on the evolution of high technology in the area and, once a year, disseminates a detailed annual community and statistical profile which is available on the centre's website. Since the study matched strongly with the mission and activities of OCRI, authorisation for the use of the OCRI database was given.

A total of 1,538 questionnaires were e-mailed. The mail-out presented the research team, the purpose of the study, ethical guiding principles, considerations on confidentiality and a hyperlink to the survey's website. The survey was deployed in five sections and provided a mix of open-ended and closed-response questions. The first section covered the profile of the respondent companies. The second section proposed four criteria and invited participants to rank three North American cities, selected from an inventory of 24, according to which they considered the best. In the third section, participants ranked those 24 cities in three categories: best, average, and worst. Finally, section four of the survey sought to discover in which of these 24 cities respondents would eventually relocate their company or establish a new branch. In the closing section, participants were asked to indicate their job title and the position held in the company.

It is worth asserting at this juncture that self-image, image and the knowledge city remain an important consideration in destination selection for both companies and employees. Knowledge workers, valued for their education and expertise, are both mobile and less likely to select living in a city with a bad image related to crime, pollution, or the hard-to-measure "X" factor of the uncool (van Winden *et al.*, 2005).

The Results

Response Rate

Out of the 1,538 survey questionnaires administered, 252 were completed and returned, thereby providing a 16% response rate. Non-respondents were contacted twice by e-mail. Several reasons explain this low response rate: a lack of interest, fear of spam, a lack of time or uncertainty as to who could or should respond to the survey (for instance, the owner or the general manager).

Participants

In the closing segment of the survey, participants had to indicate their job title. The following job titles were tabulated: 161 CEOs, 63 vice-presidents, 25 managers and 2 public relations agents. Most of the responding companies were SMEs (small and medium-sized enterprises) and this would explain why so many chairpersons and directors replied.

Profile of Respondent Companies

This section of the survey pursued two objectives: to establish a profile of the company and to search for the link between this profile and the respondent's perception on the standing of Ottawa among North American high-tech cities. Given that the sample collected turned out relatively small numbers, we were unable to assess the second objective.

Most of the responding companies were small businesses: 178 had fewer than 50 employees, 74 had more than 50 and one (1) had 270 employees. In real life, according to Navarez (2003), this nonetheless reflects North American high-tech companies. However, there are large companies in Ottawa, as shown in the following table.

Table 2. Larger High-tech Companies in Ottawa

Companies	Local employees
Nortel	5,500
Alcatel	1,890
Calian	1,763
Cognos	1,500
EDS	1,400
CGI	1,400
Anjura	1,250
IBM	1,100
Mitel	900
MDS Nordion	800
ADGA	600
JDS	550
Corel	400

Source: OCRI (2005).

Geographically, 246 of the 252 participating companies are headquartered in Ottawa. Of the remainder, three are headquartered in Gatineau, three in Montréal, one in Toronto, one in Boston and one in San Diego. 91 firms specialise in research and development (R&D), 110 perform services, 35 are manufacturers and 16 remain unspecified (as *Other*).

Looking through the cluster's grid, the results obtained show that 96 companies deliver services, 63 develop software, 23 belong to the microelectronic cluster, 28 fit in the telecommunications cluster, 14 specialise in photonics, 13 in bio-tech and 10 replied with Other. Even if the f of the sample warrants caution when comparing and combining data, the results approximate Ottawa's profile. For instance, in 2003, OCRI reported 621 businesses in the service sector, 446 in software design, 217 in the telecommunications trade, 132 in microelectronic, 66 in biotech and 56 in photonics. The foregoing includes government and private corporations.

Next, an examination of how high-tech employees perceive Ottawa and how this compares with existing statistics is addressed.

Social and Economic Competitiveness
The Three Best Cities

In each and every question in this section of the survey, participants were asked to choose from a list of 24 Canadian and North American cities (Appendix A). These twenty-four cities, all with a population in excess of 500,000, ranked in the top tier of the select ninety and were chosen for the following three characteristics: (i) the proportion of university graduates, (ii) the proportion of PhD holders, and (iii) the share of employees working in professional, scientific and technical services (see ranking in Appendix B). Among the 24 selected cities, two Canadian Census Metropolitan Areas (CMAs) were not included in the top 30: Halifax with less than 500,000 residents, and Winnipeg where hardly any high-tech activities occur.

At first, participants in the survey had to select the three best cities out of the 24 for each of the following categories: educational level, public safety, openness and vitality of the high-tech sector. Due to the limited sample, the reader must remain aware that in this and the following sections of the survey, ascertaining links between the specific playing field of the ventures and the answers provided was avoided. The size of the sample prevents the establishment of these connections and this option was waived early in the analytical process.

Education

Boston was identified by 103 participants in the study (41%) as being the city with the best-educated workforce in the local high-technology industry, followed by San Francisco (including Silicon Valley) with 79 ballots and Toronto with 45. We suggest that the proximity of Toronto and the intensive contacts with companies in Ottawa would explain Toronto's ranking. Table 3 clearly shows that participants were on the mark for Boston and San Francisco but completely off of it when it came to Toronto. Toronto actually ranks 34th in the sample of 90 North American cities of 500,000 in population for the rate of graduates per 1,000 and 54th for PhD holders. Ottawa ranks 13th in the first case and 6th in the second.

Table 3. Perception of Educational Achievement

Rank (survey)	Perception (# respondents/% votes)	Rank (stats)	Actual (grads per 1,000)	Rank (stats)	Actual (PhDs per 1,000)
1	Boston (103/41%)	4	231	4	1.15
2	San Francisco (79/31%)	1	252	3	1.29
3	Toronto (45/18%)	34	174	54	0.54

Source: Statistics Canada (1998) and NSF (2001). See *Formal education* in Appendix B.

Public Safety

The indicator used in this section involves the number of homicides per capita occurring in the community (see *Homicide rate* in Appendix B). As anticipated, Canadian cities lead in this category as they display the lowest homicide rates. Ottawa stands first with one third of the ballots (95 respondents, 38%) followed by Halifax and Québec City (at 39 each, 15%); however, the latter two cities do not belong in the list of high-tech cities under investigation. The only American city selected by respondents was Seattle (10 respondents, 4%) which ranks 18th among the 90 cities of the larger sample; Seattle displays a homicide rate of 3.1 per 1,000 while Grand Rapids, Michigan, yields the lowest homicide rate (2.1) in the roster of the 90 cities. Table 4 indicates that the perception of respondents is in line with the available data and this section testifies to the reputation of Canadian cities in terms of law and order.

Table 4. Homicides

Rank (survey)	Perception (n respondents/% votes)	Rank (stats)	Actual (hom. rate/1,000)
1	Ottawa (95 resp. / 38%)	1	0.4
2	Halifax (39 / 15%)	Not incl.	Not incl.
3	Québec City (39 / 15%)	2	0.7

Source: Statistics Canada (2001) and FBI (2004).

Access to Market

In terms of access to the marketplace, New York comes first with 98 votes (39%) followed by San Francisco (66, 26%), and Boston (49, 19%). Toronto ranks 5th with 25 votes (10%). The reader is reminded that this study is based on the perception of business people in Ottawa and it is interesting to notice how New York is well perceived although the high-tech sector is not outstanding there. The message conveyed is that New York is deemed an excellent business centre with efficient transportation and accessibility; New York offers tremendous opportunities in networking and advertising. While Toronto is rather well perceived in terms of access to market, Ottawa (4, <1%) and Montréal (2, <1%) are not recognised in this respect. Table 5 shows the percentage of direct incoming international flights used as the indicator (there are others) for access to the marketplace in this section of the survey. Ottawa business leaders have a good perception of the top 3 cities in this category except for Toronto which ranks 18th.

Table 5. Direct International Incoming Flights

Rank (survey)	Perception (n respondents/% votes)	Rank (stats)	Actual (% direct int'l inc. flights)
1	New York (98 resp. /39%)	2	60%
2	San Francisco (66 /26%)	6	29%
3	Boston (49 /19%)	5	35%

Source: Statistics Canada (2000) and US Bureau of Transportation (2000). See *International Incoming Flights (%)* in Appendix B.

High-technology Vitality

Not surprisingly, participants in the survey favourably perceive the vitality of Ottawa's high-tech business community. As shown in Table 6, Ottawa ranks second (81 ballots, 32%), between San Francisco (1st with 113 votes, 45%) and Boston (3rd with 69 votes, 27%). One could think of a parochial bias, but the survey may as well evidence the buoyancy of Ottawa's local high-tech economic activity despite the problems encountered over the last four to five years. It is notable that Toronto (25 ballots), Montréal (15), and Vancouver (13) are not equally well perceived. In order to compare the results of the survey with actual data or indicators, the proportion of jobs in the professional, scientific and technical services in existing local high-tech economies was employed. In this respect, it can be observed that the perception coincides with reported data. However, in comparison with Table 1 above, it is evident that Washington-Baltimore (1st), Raleigh, Calgary, Toronto and Vancouver rank in the top 10 but participants ignored them.

Table 6. Jobs in the Professional, Scientific and Technical Services

Rank (survey)	Perception(n respondents/% votes)	Rank (stats)	Actual (% of jobs)
1	San Francisco (113 resp./45%)	2	11%
2	Ottawa (81/32%)	4	10%
3	Boston (69/27%)	8	9%

Source: Statistics Canada (2001) and US Census (2000).

Relative Competitiveness of High-tech Cities

In the third part of the study, participants were asked to estimate the competitiveness of the 24 high-tech cities in the sample. Participants were invited to categorise each of the 24 cities under one of three headings as follows for competitiveness: best, average, and worst (Table 7). Given the subjectivity conveyed in this particular component of the process, the results are not compared to statistics. In the *best* category, San Francisco tops the bill for competitiveness (97 votes, 38%) followed by Boston (79, 31%) and Toronto (60, 24%).

Incidentally, Ottawa (57, 23%) and New York (55, 22%) fare quite well, which is a surprise at least in the case of Ottawa. Nevertheless, how could Toronto, Ottawa and New York rank better than Raleigh's (45, 18%) famous research triangle? Toronto's good standing may be explained by (i) the proximity (less than an hour's flight from Ottawa) and (ii) the presence of Torontonian ventures (e.g., Nortel) that maintain R&D facilities in Ottawa. Only five participants have included Montréal (a 2-hour drive from Ottawa) with the best. Finally it should be noted that Canadian cities are somewhat under-represented in the *best* category for competitiveness: Calgary (20, 8%), Montréal (15), Edmonton (3), Halifax (2), Winnipeg (2) and Québec City (1).

Table 7. Competitiveness among High-tech Cities

Best (n respondents/%)	Average (n respondents/%)	Worst (n respondents/%)
San Francisco (97 resp./38%)	Washington-Baltimore (57 resp./23%)	Québec City and Winnipeg (80 resp./32%)
Boston (79/31%)	Chicago (49/19%)	Halifax (49/19%)
Toronto (60/24%)	Montréal (48/19%)	Edmonton (48/9%)

In the *average* category for competitiveness, Montréal ranks 3rd with 57 ballots. Washington-Baltimore (49) prevails over Chicago (48) and Denver, while Philadelphia and Seattle (56) are on par with Montréal minus one ballot only. The *worst* category for competitiveness according to the perception of participants is populated by Canadian cities where Québec City (80 votes, 32%) and Winnipeg (70, 28%) would envy Halifax (49, 19%) and Edmonton (48). In the *worst* category, Calgary and Minneapolis share the 4th position with 28 ballots (11%);

given the vigour of their respective high-tech business sectors (Tremblay, 2002; Polèse and Tremblay, 2005), this score was unanticipated.

Relocating and Branching

In this last section, participants were asked in which city, in the sample of 24, they would relocate their business or establish a new branch. Boston (92 votes, 37%), San Francisco (61, 24%) and Toronto (47, 19%) rank at the top of the list. Vancouver was chosen by 26 participants (10%) which is a surprise because (i) it is the first time this Canadian West Coast city ranked so well and because (ii) Vancouver is not a major player in the category of high-tech cities. Natural surroundings, proximity to Whistler and other natural amenities, climate, and quality of life could explain this choice. Another major surprise was the poor performance of world-class high-tech cities such as Austin (27, 11%) and Raleigh (24, 9.5%). Montréal, with 17 votes (7%), is nevertheless ahead of Calgary (13, 5%), Edmonton (13), Halifax (12), Winnipeg (2) and Québec City (1).

Table 8 presents the five best cities as perceived by the participants using the selected criteria. Boston and San Francisco rank on top in four out of five criteria. Austin, a stronghold of high technology, did not rank among the top 5 and Raleigh comes into sight only twice. Since Ottawa and Toronto show up most often in the five categories, it is suggested that for the participants, Ottawa does belong with the outstanding North American high-tech cities. We notice that the cities appearing in Table 8 also show up in Table 6 (employment in the professional, scientific and technical services in Canada and the US). This outcome would suggest that participants share the cultural reality of North American high-tech cities. Finally, concordance of results in both Tables 8 and 6 tend to support other statistical results obtained in Polèse and Tremblay (2005), all of which converge to further strengthen the perception of the local business community regarding Ottawa's place among North American technopoles.

Table 8. Summary of the Survey Results
The Five Best by Category

Safety	Education	Access to Market	Vitality	Competitive-ness	Relocation or Branching
Ottawa	Boston	New York	San Francisco	San Francisco	Boston
Halifax	San Francisco	San Francisco	*Ottawa*	Boston	San Francisco
Québec City	*Toronto*	Boston	Boston	*Toronto*	*Toronto*
Calgary	*Ottawa*	Los Angeles	Raleigh	*Ottawa*	*Vancouver*
Toronto	Raleigh	*Toronto*	Atlanta	New York	San Diego

Conclusions

The results of the inquiry clearly reflect that Ottawa's high-tech business community maintains an excellent opinion about their city and region, even in the aftermath of the massive layoffs in 2001. Ottawa's standing in terms of (i) public safety, (ii) availability of a highly skilled work force, (iii) local vitality and (iv) competitiveness of high-tech companies stand in comparison with other North American high-tech clusters and is likely to compensate for a perceived weakness in terms of (v) access to market.

In another study by Polèse and Shearmur (2004) with a shift of focus that looked at Montréal rather than Ottawa, Montréal was compared with eleven North American cities on the basis of eight criteria about competitiveness alone. In the 2004 study, Ottawa still ranked first (1 occurrence), second (3 occurrences) and fourth (2 occurrences), a remarkable score in six of the eight indicators.

Comparative studies are unfortunately infrequent. This investigation, including Polèse and Shearmur (2004), take into account both the perception of business leaders and the dimensions sketched out in official statistics. Further studies could be carried out in the US to measure the American perception of Canadian high-tech cities; urban marketers and business communities would benefit from the knowledge generated in such investigations that discriminate or channel and temper allegations of standing and calls on the basis of statistical analysis.

References

Centre on Governance. (2002). *Ottawa works: a mosaic of Ottawa's economic and workforce landscape. Report 1: Ottawa's workforce environment.* Ottawa: Human Resource Development Canada.

De la Mothe, J., Paquet, G. (1999). *Information, innovation, and impacts.* Boston: Kluwer.

Fache, J. (2002). *Les territoires de la haute technologie.* Rennes: Presses universitaires de Rennes.

Holzman, J. (1994). *An economic profile: Ottawa.* Ottawa: Office of the Mayor, City of Ottawa.

ICF Consulting. (2000). *Choosing a future: a new economic vision for Ottawa: final report.* Ottawa: Ontario Ministry of Economic Development and Trade, Office for Urban Economic Development.

Mallett, J. (2002). *Silicon valley north: the formation of the Ottawa innovation cluster.* Ottawa: Information Technology Alliance of Canada.

Miller, S., Adams, S. (2001). *Defining the British Columbia high technology sector using NAICS.* Victoria: BC Stats and Ministry of Competition, Science and Enterprise.

Navarez, L. (2003). *New money. nice town*. London: Routledge.

OCRI (Ottawa Centre for Research and Innovation). (2003). *Ottawa 2003: technology industry guide*. Ottawa: OCRI.

Polèse, M., Shearmur, R. (2004). *Le positionnement de Montréal par rapport à 11 autres agglomérations: entre perceptions et statistiques*. Montréal: INRS-UCS.

Polèse, M., Tremblay, R. (2005). L'économie du savoir et la manie des "rankings": une analyse intégrée des villes canadiennes et américaines. *Le géographe canadien/Canadian Geographer*, *49* (2), 198-213.

Shavinina, L.V. (2004) *Silicon valley north. A high-tech cluster of innovation and entrepreneurship*. Amsterdam: Elsevier.

Statistics Canada. (2005). Experienced labour force 15 years and older by industry, by census metropolitan areas (2001 Census). Statistics Canada. Retrieved on May 14, 2006 from http://www40.statcan.ca/l01/cst01/ labor47e.htm?sdi=experienced% 20labour%20industry.

Statistics Canada. (2001). *Ottawa-Hull community profile*. Statistics Canada Retrieved on May 14, 2006 from www.statcan.ca.

Tremblay, R. (2002). *La passion pour les classements: revue des principaux "rankings" des technopoles américaines et canadiennes.* Montréal: INRS-UCS.

Tremblay, R. (2003). Le tandem Gatineau-Ottawa ou la frontière fantôme. In Coté, R. et Venne, M. (eds.). *L'annuaire du Québec*. Montréal: Fides, 240-248.

Tremblay, R. (2004). La place d'Ottawa-Gatineau parmi les villes du savoir nord-américaines. *Organisations et territories*, *12* (3), 12-20.

Tremblay, D.-G., Rolland, D. (2003). *La nouvelle économie. Où? Quoi? Comment?* Québec: Presses de l'Université du Québec.

van Winden, W., van den Berg, L., Pol, P. (2005). European cities in the knowledge economy: towards a typology. *Urban Studies*, *44* (3), 525-549.

APPENDIX A

Table 9. Rank by Population – The Sample of 24 Cities

Cities	Population
New York	21,199,865
Los Angeles	16,373,645
Chicago	9,157,540
Washington-Baltimore	7,608,070
San Francisco	7,039,362
Philadelphia	6,188,463
Boston	5,819,101
Dallas	5,221,801
Toronto	*4,907,000*
Atlanta	4,112,198
Seattle	3,554,760
Montréal	*3,551,100*
Minneapolis	2,968,806
San Diego	2,813,833
Denver	2,581,506
Vancouver	*2,009,940*
Austin	1,249,763
Raleigh	1,187,941
Ottawa	*1,063,665*
Calgary	*969,600*
Edmonton	*954,100*
Québec City	*694,000*
Winnipeg	*684,300*
Halifax	*359,183*

Source: US Census (2000) and Statistics Canada (2001).

APPENDIX B

Table 10. The 20 Finest North American Cities with more than 0.5M Population

Rank	Homicide rate (lowest on top)	Bachelor Degree holders n/1,000	PhD Degree holders n/1,000	% jobs in prof., sci. & tech. services (highest on top)	% international air traffic (highest on top)
1	Cleveland	San Francisco	Raleigh	Washington-Baltimore	Miami
2	Kansas City	Raleigh	Washington-Baltimore	San Fran-cisco	New York
3	Charlotte	Washington-Baltimore	San Francisco	*Calgary*	Honolulu
4	Greenville	Boston	Boston	*Ottawa*	Los Angeles
5	Scranton	Denver	Austin	*Toronto*	Boston
6	Youngston	Austin	*Ottawa*	Raleigh	San Francisco
7	Charleston	Minneapolis	Albuquerque	Denver	Houston
8	*Ottawa*	Seattle	Tucson	Boston	Chicago
9	*Québec City*	Atlanta	Albany	Albuquerque	Detroit
10	*Calgary*	New York	San Diego	Vancouver	Philadelphia
11	*Toronto*	West Palm	Springfield	San Diego	*Vancouver*
12	*Hamilton*	Hartford	Knoxville	Austin	Minneapolis
13	*Vancouver*	*Ottawa*	Denver	New York	*Montréal*
14	Grand Rapids	Colorado Sp.	Philadelphia	Atlanta	Seattle
15	*Montréal*	Richmond	New York	*Montréal*	Dallas
16	Honolulu	Jacksonville	Columbus	Colorado Sp	Orlando
17	Portland	Albany	Hartford	Minneapolis	Atlanta
18	Albany	Sarasota	Seattle	Chicago	*Toronto*
19	Boston	San Diego	Columbia	Houston	Greenville
20	Minneapolis	Columbus	Bâton Rouge	Philadelphia	Washington-Baltimore

Source: Polèse and Tremblay (2005).

APPENDIX C

Note 1: Homicide rate

Rate of first and second-degree murders /1,000 population, voluntary homicides, involuntary homicides and infants murdered. American data are from the FBI and should be specifically applied to the Metropolitan Statistical Area (MSA) level. FBI data is for the year 2001. Chicago and St. Louis results are lacking due to a different territory definition. Albany, Dallas and Little Rock results are from 2000 census while those for Indianapolis are dated 1999. Canadian data was extracted from the Statistic Canada Census in 2001. The rate for Ottawa-Gatineau excludes Gatineau.

Note 2: Formal education

Data on education are extracted from US Census 2000 and Statistics Canada 2001. They apply to people over 15 years old with a first cycle grade or higher. The rate /1,000 is divided by the total population including those who are 15 years old and younger.

Note 3: International incoming flights (%)

This informs the reader on the percentage of international direct flights/total air traffic in all airports in a city. For example, in Montréal, data includes both the Trudeau and Mirabel airports. American results are extracted from US Bureau of Transportation Statistics (2000). They do not include Canadian and Mexican destinations. Canadian results are extracted from the Statistics Canada Census (2000) and do not include direct flights to United States. Notice that Hartford, CT and Springfield, MA share the same airport facilities.

Part II

Planning the Cluster
By Decision, by Design Or by Destiny?

CHAPTER 3

Local Governments' Record of Assessing the Impacts of the High-tech Industry on Ottawa's Land Use-Transportation Relationship

1970s-2005

Barry WELLAR and Nick NOVAKOWSKI

University of Ottawa and
Memorial University of Newfoundland

Introduction[1]

For the purposes of this chapter, the high-tech industry is defined to include firms that engage in research, development, and production or distribution of goods and services in the computer, telecommunications, and emergent technology sectors. Our primary research objective is to describe and analyse the record of local governments in assessing the impacts of the high-tech industry on Ottawa's land use-transportation relationship. A secondary research objective, identified during the research design phase of the study, is to provide a preliminary evaluation of the document – based approach that was devised to conduct the investigation. As discussed below, this approach was developed upon realising that we would not have access to the data needed to empirically examine the impact assessment record of local governments.

We begin by presenting four points of context to provide the rationale behind the structure of the chapter. The context points are followed by a brief description of the high-tech industry in Ottawa. After presenting the study methodology, we proceed to the chapter's core section in

1 The editorial assistance of Marjorie Wellar is gratefully acknowledged, as is the assistance of John Tinholt in generating the Ottawa-Gatineau regional map using Geogratis data.

which we derive, and then apply, the criteria selected to examine the extent to which local governments assessed the impact of the high-tech industry on Ottawa's land use-transportation relationship from the 1970s to the present.

As the first point of context, and beginning with the water mode for lumbering, trade and defence purposes, and moving through the road, rail and air modes, the land use-transportation relationship in Ottawa has been evolving for more than 150 years. However, the high-tech industry did not make a strong appearance in the Ottawa area until the 1960s. As a result, while our chapter has a temporal window of approximately 40 years, it has to take into account what happened earlier to establish the land use-transportation relationship into which the high-tech industry entered in the 1960s.

Second, policies, plans and programmes of federal, provincial, regional and municipal governments have all affected Ottawa's land use pattern and densities, the structural and functional characteristics of Ottawa's transportation system(s) and, consequently, Ottawa's land use-transportation relationship. Their contribution to the map of Ottawa's land uses and transportation system components can be summarised as follows.

Federal government involvement in Ottawa's land use and transportation history and geography is represented through the activities of such agencies as Public Works, Transport Canada, and the National Capital Commission (NCC). It is manifested by federal buildings, parkways, the Rideau Canal, the Greenbelt, and segments of the urban greenways system, as well as the 100,000-plus federal employees who buy homes, rent apartments, and make work, shopping, institutional and recreational trips by car, bus, rail, bicycle and/or walking. As shown in Figure 1, the National Capital Region (defined as the jurisdiction of the NCC) includes most of the City of Ottawa, but not all. Further, while the map labels the new Québec-side entity of the Ville de Gatineau, it fails to delineate the full expanse of the Census Metropolitan Area (CMA) of Ottawa-Gatineau as defined by Statistics Canada (and by which demographic statistics are generated for planning, transportation, housing, and other studies). These map features are noted should questions arise about the comparability of our study area with those of other contributors.

Figure 1. The Ottawa-Gatineau CMA in 2006

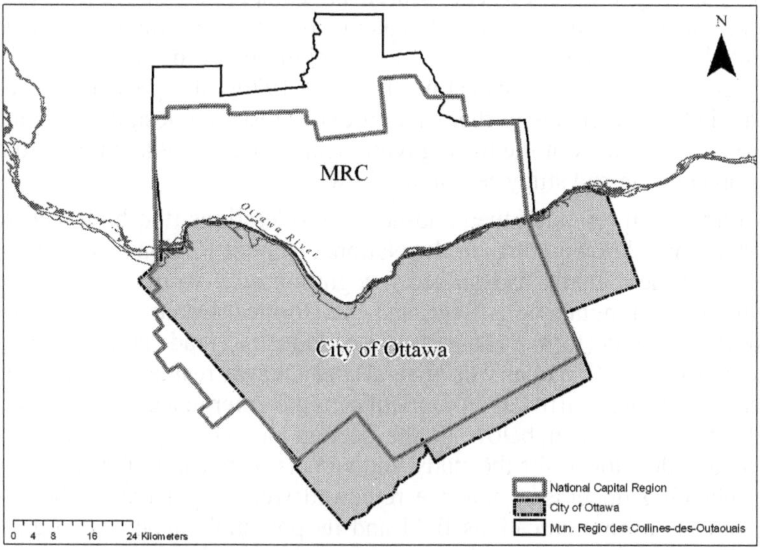

City of Ottawa - Regional Context

Source: John Tinholt, 2007.

At the local government level, as many as 12 municipalities and a regional body made decisions on land use and transportation matters until January 2001, when amalgamation took effect and the (new) single-tier City of Ottawa was created. Through official plans, zoning by-laws, transportation master plans and other instruments, these local governments were and are responsible for deciding which non-federal lands are used for what purpose, and the transportation connections between those lands.

Third, while debates and decisions about where to locate high-tech firms came to the fore in the 1980s and 1990s, it is necessary to examine planning processes from a number of years earlier. That condition applies because development activities in 1995, 2000 or 2005 could be shaped by planning decision made in 1965, 1975, 1985, and so on.

As discussed below in the Study Methodology section, the approach taken to conduct this research is to use planning documents produced in the 1960s and 1970s as the basis for examining local governments' record of assessing the impacts of the high-tech industry on the land use-transportation relationship. These documents contain many of the normative statements that characterise current land use and transportation goals, objectives and principles, so we are comparing 'apples to apples' insofar as the then-and-now aspects of land use and transportation planning are concerned.[2] Moreover, comparison of local government vision statements, mission statements, and other planning reports printed over the decades reveals a *very* high degree of similarity. This degree of association connecting the early 1970s, the present, and the years in between lends further support to our research design decision to take as long a look at the local government impact assessment record as documentation and study resources allow.

Further, in terms of then-and-now considerations, the high-tech industry was a known entity in the National Capital Region in the 1970s. By that time Canada had gained recognition as a world leader in the computer/communications field, and the (former) Department of Communications and (former) Department of Industry, Trade and Commerce had introduced Canadians in general and Ottawa residents, firms, and local government officials in particular to the emergence of a high-tech industry.[3] As shown below in the section on the impact assessment approach developed for the study, individuals responsible for producing the planning documents that we reviewed were cognizant of the burgeoning telecommunications field and its potential capacity to significantly modify Ottawa's land use-transportation relationship.

Fourth, there is the matter of establishing whether our study has already been done or, if not, if it has been done in part. In the former case there might be no need for this chapter, and in the latter circumstance it would be necessary to acknowledge any related work and incorporate as appropriate.

Our review of the extant literature revealed that the former Region of Ottawa-Carleton, the former cities of Ottawa, Nepean and Gloucester, and the current City of Ottawa funded numerous land use, transportation and development studies, conferences, visioning exercises, and public

---

2 Informal tests of the validity of a planning document-based approach for impact assessments are provided by the newspaper search and review procedure used by B. Wellar while at the Ministry of State for Urban Affairs in the 1970s (Wellar, 1975), in research and courses in planning at the University of Ottawa 1979-2004 (Wellar, 1981,1990, 2000), and for a commissioned report on the Ontario Municipal Board (Wellar, 2004).

3 B. Wellar was Senior Research Officer and Senior Policy Advisor at the Federal Ministry of State for Urban Affairs in the 1970s, and frequently served as a delegate on Interdepartmental and OECD Panels on computers/communications and information technology. The high regard for and understanding of Canadian contributions to the emerging 'information society' were witnessed on numerous occasions at the international, national, and local levels, and were seen in the 'press packages' which reported on high tech articles, editorials, letters, stories etc. that frequently appeared in area newspapers as well as on area radio and television broadcasts.

consultation projects over the past three decades, many of which had price tags in the range of 200,000 to 300,000 Cdn$. Indeed, several of the initiatives that involved international participants, economic development concerns, large-scale transportation issues, or extended study periods were considerably more costly and had million dollar budgets. However, examination of reports from these policy, planning, development, and research activities hosted by the area's local governments did not yield any evidence of an empirical assessment of the impact of the high-tech industry on Ottawa's land use-transportation relationship.

Further, the review of the published reports did not yield any non-trivial additions to the knowledge base about Ottawa's land use and transportation 'systems,' nor of the evolving land use-transportation relationship. And, to complete the literature review cycle, our review did not identify any contributions made by those projects and studies to the theoretical or applied research methodology underlying land use and/or transportation studies.[4]

This brings us to the secondary research objective. As noted above, this element of the chapter was added as a result of learning that although area municipalities have sponsored numerous expensive, topically-related studies and projects over several decades, contributions to impact assessment information and methodology were extremely limited in scope, depth and utility. In response to the findings about the limitations of research, visioning, etc., done to date, therefore, and the apparent likelihood that the pattern would continue unless an alternative is presented, this chapter was re-designed to take on an added research task. That is, as a contribution towards 'raising the research bar' in Ottawa and elsewhere, the chapter includes an evaluation of whether our approach of 'revisiting' statements in official plans and associated documents serves as an effective, pertinent means to examine local governments' record of assessing the impact of the high-tech industry on Ottawa's land use-transportation relationship.[5]

---

4 A newspaper search and review was also undertaken in anticipation that a consultant, visiting expert or member of staff of a local government might offer a relevant observation in regard to the impact assessment topic. However, analysis of numerous columns, editorials, articles, letters, etc., that referred to or discussed the projects and reports also failed to yield any findings or insights that could contribute to the design of this paper.

5 The larger question of the *general* validity of this approach as a means of achieving both substantive subject matter and methodological contributions to current planning, transportation, and impact assessment studies is a topic for another paper.

Introduction to Ottawa and its High-tech Industry

The Ottawa-Gatineau census metropolitan area (CMA) had a population of 1,063,664 in 2001, and since 1996 experienced a growth rate of 6.5% – well ahead of the national average of 4% (Statistics Canada, 2004a). Within the CMA, the Ontario side experienced an even higher rate of growth at 7.2%, yielding an increase in Ottawa's population from 751,646 in 1996 to 806,096 in 2001 (Statistics Canada, 2004b; 2004c). Median family and per capita incomes in Ottawa exceed those in all of Canada's major cities, as do levels of educational attainment and per capita green space (Statistics Canada, 2004d).

As stated in the Introduction, the National Capital Region (NCR) roughly approximates the boundaries of the Ottawa-Gatineau CMA. The city's jurisdictional complexity is compounded by the presence of the federal government as both a decision maker affecting local land use and transportation decisions, and the region's dominant employer. In 2003, there were 112,310 federal employees in Ottawa-Gatineau, which represents an annual increase of 3,000 to 6,000 employees per year since 1999 (Statistics Canada, 2004e).

At the peak of the high-technology boom in 2000, high-tech employment reached 85,000 and for the first time exceeded federal employment on the Ottawa side of the Ottawa-Gatineau CMA. However, more than 20,000 technology-related jobs were lost following the stock market crash and high-tech sector downturn over the summer of 2001, leaving only 64,500 high-tech employees in 2003 (Harrison, Cooper and Mason, 2004). Once again, the federal government became the City of Ottawa's dominant employer. As of this writing, high-tech employment has not recovered to levels experienced in 2000, but it remains the area's major non-governmental employment sector.

In addition to numbers of employees being of importance to this study, the structure of the high-tech sector is also pertinent. That is, there are three basic types of high-tech company, and each type provides or offers different products or services, which could be cause for them to have different impacts on Ottawa's land use-transportation relationship. Since other chapters discuss these differences in detail, we focus our remarks on the transportation aspect and mention illustrative technologies and companies.

- *High-tech manufacturing*. These firms produce computers, high-tech equipment (aeronautics, medical technologies), peripherals and components. Their transportation preference is for superior road, rail, air and sea connections in order to move their products to their markets.

- *Communications services*. These firms sell equipment or products (computers, cell phones, satellite TV), and/or provide support or services (call centres, Internet Service Providers (ISPs)) for equipment purchasers. They require superior infrastructure including trunk phone lines, fibre optic cables and satellite systems. A location close to markets is vital for ISPs due to long distance charges, but not for call centres using international 1-800 numbers.
- *Technology-related services*. Characterised by such higher-end functions as software design and development, and computer training, this type of firm wants to be part of a local networking of businesses and expertise that is clustered.

Over the history of the high-tech industry in Ottawa, each of these types of enterprise has been well-represented. However, the movement of companies into the Ottawa area to join the existing sets of high-tech firms has been relatively limited. Rather, the high-tech industry is considered 'home-grown,' with its origins primarily related to the presence of federal government agencies (National Research Council, Department of National Defence) and the associated spin-offs involving each type of high-tech firm.

The first wave of start-ups began in the 1950s, and focused on defence-related technologies. In the 1970s and 1980s, the telecommunications start-ups began (Mitel) along with computing equipment (Gandalf Technologies, Norpak), lasers (Lumonics) and software (Cognos). In the late 1990s, biomedical and telecom-related development began to emerge in photonics, optical networking, and wireless technologies (Wolfe, 2002; Harrison, Cooper and Mason, 2004). As indicated, while Ottawa's high-technology industry is dominated by information and communication technology (ICT) firms, these enterprises are only part of the city's high-tech industry profile. Semi-conductor development, software design, and the emerging photonics sub-sector are also important elements, as are the emerging sub-sectors in the life sciences, biotechnology and biomedical devices (Wolfe, 2002).

By way of comparison, and accepting that the definition of high-tech employment varies from country to country and agency to agency, it appears to be widely recognised that the only area in North America with a higher per capita rate of employment in high technology is Silicon Valley in California (Ottawa Business Journal, 2004). Moreover, and of considerable significance to the land use-transportation relationship and the research objectives of this chapter, more than 90% of Canada's R&D in telecommunications is undertaken in the Ottawa region (Harrison, Cooper and Mason, 2004).

In summary, the high-tech industry has emerged as a major force in Ottawa's economy, and the presence of firms from each of the three basic types of high-tech company in principle makes the industry's impact on Ottawa's land use-transportation relationship a matter of major policy, planning and development significance.

Study Methodology

In this section, we establish the rationale and validity of the approach taken to conduct the impact assessment (IA) review. As a brief note of context, IAs are longitudinal rather than cross-sectional studies, and are based on datasets that permit ascertaining the direction and strength of multivariate relationships over extended lengths of time and multiple geographic areas or regions. However, and as discussed in previous reports on the technical capabilities of local governments to undertake or contribute to IAs (Wellar 1981; 1996; 2003), few cities in Canada can support that kind of robust study for even the past year, much less for the past decade or, even more unlikely, for the past two, three or four decades.

Regrettably, for the purposes of this chapter, local governments in Ottawa-Carleton were no exception to the rule. Despite all the studies, summits, projects, etc., which have been funded over the years, no report describing a multivariate, longitudinal, quantitative impact assessment was located, nor did we encounter any indication that such an assessment of the impact of the high-tech industry on Ottawa's land use-transportation relationship was under consideration. As a result, we could not use the inputs and outputs of the traditional IA approach to establish and examine the IA record of local governments regarding high technology and the land use-transportation relationship. And, we were precluded from undertaking even a partial, empirical IA study due to lack of access to needed data.

That combined shortcoming – no prior IA studies by local governments and no IA database – meant that the original research design needed to be replaced by an alternative IA approach. However, a search of the literature did not yield any tested precedents that we could follow, nor any advice on devising a 'Plan B' for examining local governments' IA record.

Fortunately, our familiarity with the planning documentation on Ottawa enabled us to identify an alternative approach for this impact assessment review. The approach involves 'back-tracking,' and is based on identifying statements made in local government documents some 20 or more years ago which are totally or largely consistent with statements made over the intervening years, and are made in recent or current planning documents. These statements can represent goals, objectives, principles or practices. Their key feature is that they allow us to ascertain whether and how local governments had regard for the impact of the high-tech sector on the land use-transportation relationship *over time and space* in the City of Ottawa.[6]

Upon acquiring that information, an assessment can then be made as to the degree and extent that the land use-transportation relationship was affected by the high-tech industry. This activity includes identifying changes that the sector caused to be made to Ottawa's land use and transportation goals, objectives, principles or practices and, consequently, to the land use-transportation relationship.

The next section outlines the IA approach designed for this chapter. It bears emphasising that this is not a funded study. As a result, while our research design adopts a broad scope when setting out the study parameters, we are obliged to be selective in terms of the amount of detail that we can provide and the intensity of our analysis.

The Impact Assessment Approach

In the next several pages, we present sections from several planning reports in order to establish the rationale and validity of our approach. As indicated, the statements make a substantive case for using this documentation as the foundation for ascertaining and evaluating the local government record of assessing the impact of the high-tech industry on the land use-transportation relationship. The italics below are added to draw attention to terms and phrases that are directly pertinent to the research objectives of this chapter.

> This report on travel forecasts establishes the future travel requirements of the first two development stages of the Region's draft Official Plan. It is therefore concerned with the *inter-relationships between land use and transportation*, the forecasting and analysis of future travel, and the establishment of the balance between the various components of the transportation network in accommodating this travel. The next report is a preliminary examination of the implications of rapid transit in the Ottawa-Carleton Region in terms of possible routes, station locations, system characteristics, *impact on development and development options* (RMOC, 1973, p. i).

---

6 While elaboration of this observation is beyond the scope of the present paper, a preliminary examination suggests that this approach lends itself to other types of impact assessments for Ottawa, and it may also be applicable to other Canadian municipalities.

The *existing transportation system is definitely inadequate* to meet this three-fold increase in travel demand (RMOC, 1973, p. iii).

Not only is the existing road system limited in terms of high capacity 4-6 lane facilities, but also is characterised by lack of continuity and inadequate crossing across travel barriers such as the Ottawa and Rideau Rivers [*sic*]. The Queensway is the only continuous high capacity east-west facility. No continuous high capacity facility, even of an arterial nature, exists in the north-south direction (RMOC, 1973, p. iii).

The existing public transportation system in Ottawa-Carleton is limited to a conventional bus system and a few express routes operated during peak periods…A recent survey indicated that less than 10 percent of person trips in an average week day in 1972 used public transportation in the study area…Therefore, under existing conditions the road system is carrying over 90% of daily person trips. In order *for this system to carry the projected travel at the present level of congestion,* transit must carry a minimum of 63% of peak hour travel *throughout the study area* (RMOC, 1973, p. iv).

Such levels are possible to achieve in certain locations provided that the transit system is substantially improved, and the public not only supports the improvements but also uses the provided services in preference to private transportation. However, *the attainment of this extent of transit usage everywhere in the region is impractical.* For *public transportation* to attract high proportions of trips throughout the study area and throughout the day, it *has to extend its improved service* far beyond what is envisaged and its exclusive rights-of-way have *to penetrate most neighbourhoods and communities….Travel projections and analysis* made in this report suggest the need for improvements of both public and private transportation facilities. The study indicates substantial road requirements are in the east-west direction south of the Base Line [*sic*]/Heron Road, and across the Ottawa River to the east and west of the Region. This is *because the projected population and employment distributions result in substantially high travel demands in these locations* (RMOC, 1973, p. v).

The distribution of future population and employment is fundamental to any plan for the Region. In many previous studies, these distributions have been prepared as relatively fixed forecasts which must then be served by transportation facilities, water, sewer and other services. In practice, this approach does not permit changes to be made when major problems arise at a later stage in the planning process. To overcome this limitation, *the approach taken in preparing a plan for the Ottawa-Carleton Region is to assume a number of possible future population and employment distributions whose implications for public services will be tested* (RMOC, 1972, p. 83).

There are two basic options for the location of jobs. One is *concentrated employment* within the Greenbelt; the other is more *dispersed employment* with a correspondingly higher number of job opportunities in the *specified growth areas.* Similarly, there are two alternatives for population. One is a

> concentrated, dense development within the Greenbelt; the other is to have fewer people located within the Greenbelt and the population outside concentrated in growth areas (RMOC, 1972, p. 83).
>
> Analysis of present growth patterns indicates that the population in the National Capital Region may reach 1,400,000 persons within 25 to 40 years time, depending on the rates of growth in the local economy, attitudes to family size and the degree of migration into and out of the Region. When this population is reached, it is estimated there will be 1,000,000 persons living and 450,000 working in the Regional Municipality of Ottawa-Carleton…*There are many ways to distribute this future growth.* The preliminary evaluation of alternative concepts for urban growth in the Ottawa-Carleton Region indicated that either satellite towns or urban corridors outside, but adjacent to, the Greenbelt represented the most satisfactory way of accommodating future urban development (RMOC, 1972, p. 84).
>
> Further, since many of the land-use controls in the region are weak, it is simply not practical to apply restrictive policies to existing developments, because they will merely result in a greater dispersal of such activities than already exists (RMOC, 1972, p. 136).
>
> Consideration as to what could be achieved through changes in behaviour compels the realization that *traffic congestion is largely an artificially created phenomenon.* The human race has elected to behave in a certain manner, and traffic congestion is a direct result of this decision. Similarly, the introduction of a high-capacity transit system is intended to reduce congestion, but it appears to induce even greater development, which in turns [*sic*] leads to just as much and often more congestion than existed previously…. If the urban environment is truly valued, it should be accepted that there is a definite limit to the number of people who can be comfortably and conveniently moved in any area, whether on a local or a regional scale, without significant damage to the urban environment. *It should be an important issue as to whether a high-capacity transit system that will stimulate this kind of high-density redevelopment and increased congestion is really to be desired. Perhaps greater thought should be given to the capability of telecommunications to decentralise employment*, thereby permitting the development of new towns, or the revitalization of towns that are now dying while just a few metropolitan areas in Canada continue to grow dramatically (RMOC, 1972, p. 140).

As these excerpts and numerous other statements in the planning documents make clear, it was known in the early 1970s that Ottawa was already being challenged by a difficult land use/transportation relationship. And, it was also recognised 30 years ago that significant public interventions – policies, plans, programmes, as well as basic changes in the attitudes of individuals and corporations – would be required to prevent a worsening of the land use-transportation relationship.

Further, and this is a key methodological point, the essence of those excerpts appears in City of Ottawa documents produced since amalgamation in 2001. Moreover, to further confirm their continuity over time, similar-to-identical statements appeared in reports published by the Region of Ottawa-Carleton, the City of Ottawa, and other local municipalities in the 1980s and 1990s. Therefore, it is our position that the reports provide a pertinent, representative body of documentation upon which to base this impact assessment review.

It is our further contention that these background documents also provide the means for using the planning issues, goals, objectives and principles statements of 30 years ago to conduct this assessment. That is, a normative approach is used in the planning field to specify what land uses should go where, when, and to specify how those land uses should relate to and be connected by the transportation system. Therefore, it follows that if there are a sufficient number of pertinent, representative normative statements in the 'official' literature, then it is appropriate to use that documentation as a basis for examining the local government record of assessing the impacts of the high-tech industry on Ottawa's land use-transportation relationship.[7]

Before examining the IA record of local governments, one more step is required to link the planning documents of the 1970s with those of today. This may be referred to as passing the 'relevance test.' The purposes of this test include ensuring that the matter under study is actually of public interest, that the study premises are sound, and, in particular, that the rationale used to justify the comparative study approach is sound.

By way of brief comment on the relevance test (for a detailed discussion of this topic see Wellar, 1998a), if circumstances have changed over the intervening time period then the impact assessment – related concerns of three decades ago may no longer exist. Or, they could be different. Application of a relevance test serves to ensure that we are in fact comparing planning apples to planning apples in our examination of

---

7 Readers familiar with official plan (OP) reviews will appreciate that what is proposed here is identical to that process, but with a twist. Unlike the typical OP review that only goes back five years or so, we are taking into account what was written 30 or more years ago. And, readers familiar with cumulative impact assessment procedures will also appreciate the argument made. That is, in robust IAs which recognise connectedness among system events and processes over time and space (Wellar and Novakowski, 2000), it is logically reasonable to use the 30-year timeframe in the urban domain, and especially when the research problem involves 30-40 years of industrial activity and a land use-transportation relationship that has been evolving for more than 150 years.

local governments' record of assessing the impacts of the high-tech industry on Ottawa's land use-transportation relationship.

One element of the relevance test applied to this study's research design has its origins in the popular literature; that is, newspapers. The thesis to be tested in this situation involves a three-stage process, and begins with ascertaining whether it is possible to locate articles, letters, columns or editorials regarding the 'coulds' and 'shoulds' of planning that were published in daily and community newspapers over the years, and which are relevant today. Further, to be specific to this chapter, the items would be comparable in content to current newspaper items regarding the impact of the high-tech industry on Ottawa's land use-transportation relationship. And more importantly for the methodology of this then-and-now study, the items would also provide insight and direction into the scope, depth, and regard which current City of Ottawa official documents should give to assessing the impact of the high-tech industry on Ottawa's land use-transportation relationship.

If all those conditions are met, then it may be concluded that the impact topic is of abiding public interest, that use of the long-term perspective is appropriate, and that our comparative analysis is on methodologically-sound ground. In the next section, we explore this three-stage thesis in terms of its implications for this study in particular, but with an eye to potential applications in other situations and locales.

Contribution of Newspapers to Local Governments' Record of Assessing the Impacts of the High-tech Industry on Ottawa's Land Use-Transportation System Relationship: 1970s to 2005

The role of newspapers as an important source of documentation on public discourse in Ottawa-Carleton is demonstrated by the report, *Newspapers as a Source of Fact and Opinion on Pedestrians' Safety, Comfort, Convenience: A Keyword-Based Literature Search and Review* (Wellar, 2000). That work was undertaken as part of the Walking Security Index (WSI) project, and represents one of the more complete guides on why and how to document and differentiate among the contributions that elected officials, professional staff and citizens make to public policies, plans, programmes and projects through newspaper articles. Following the framework of that model, a selection of articles typifying newspaper reportage on Ottawa's high-tech, land use and transportation evolution over the past 25 or so years is presented in Table 1.

Based on a review of extensive clippings files that begin with entries from the early 1970s, and a search of library holdings at the University of Ottawa and the Ottawa Public Library, it appears that well in excess of 30,000 items – editorials, columns, features, news stories, public meeting announcements, letters to the editor – have been published on high-tech, land use and transportation topics by area newspapers – dailies, weeklies, bi-weeklies, monthlies – since the early 1970s. While the 30,000+ figure may appear impressive, the volume of articles alone is not sufficient to establish the importance of the high-tech impacts to the evolution of Ottawa, nor to necessarily cause the impacts to receive the considered attention of local governments. However, if the articles also involve *people of influence*, then the reportage could be construed to have the authority needed to command the attention of local governments charged with addressing impact-related matters. We believe the following observations make that case.

The article titles in Table 1 typify many of the approximately 900 articles reviewed for this study. That is, there are strong similarities in terms of issues, involved parties, circumstances, perspectives, lines of argument, etc. Examination of the articles presented in Table 1 reveals that the following groups of people of influence have made frequent, substantive contributions to this area's public discourse on the impacts of the high-tech industry on the land use-transportation system relationship:

- Municipal, regional, provincial and federal politicians;
- Local government staff in planning, economic development, transportation, engineering, finance, and infrastructure development;
- Academics and researchers from the Ottawa region, as well as from across Canada and abroad, in such fields as urban and regional planning, urban geography, economics, public administration, transportation, political science, transportation planning, computers/communications, and engineering;
- Owners and managers of high-tech firms, development companies, engineering firms, building contractors, and all manner of consultants;
- Newspaper editorialists and columnists;
- Community associations and interest groups from across the region; and, most importantly,
- Ordinary citizens who elect politicians, pay taxes, and enjoy or suffer the consequences of decisions and actions that affect and are affected by the impacts of the high-tech industry on Ottawa's land use-transportation system relationship.

In addition to the quantity of articles and the involvement of people of influence in the discourse, there is the central matter of article content. As a reading of the selected articles reveals, while our primary focus is on articles that make direct or indirect reference(s) to impacts, the articles also have regard for the *contexts* in which impacts are presented, discussed, assessed and, ultimately, addressed (or not) by local governments.

Table 1. Newspaper Articles Typifying the Public Discourse on Ottawa's High-tech, Land Use and Transportation Issues, Visions, and Challenges: 1970s to 2005

Newspaper Article Title	Yr.	Nwsp *	Topic(s)**
Taking steps towards the end of the automobile era	75	A	1,2,3,4,5,6
Our capital of tomorrow	79	A	1,2,4,5
Ottawa seeks high-tech firms	80	A	4
The transitway: Has it failed to 're-shape the urban structure'?	85	A	1,2,3,5,6
Comparing notes: Region looks south for ways to cope with	88	A	2,4
The lowdown on the hi-tech sector	90	A	4
Kanata expansion battle key to regional future	94	A	1,2,3,4,6,7
Stopping urban sprawl – Intensification in the downtown core…	95	A	1,2,3,4
Urban sprawl: Other cities show us the dangers of uncontrolled development	96	A	1,2,3,4,5,6
Regional plan steers growth to inner city	97	A	1,2,3,4,5,6,7
Board warns area too reliant on technology	00	A	4
High-tech boom faces roadblocks	00	A	4,5,6
High-tech boom might crash	00	A	4
Snow crews can't keep up with area's growth	00	A	1,2,3,4,5,6
South Nepean's old roads can't carry high-tech expansion	00	A	2,4,5,6
Everything's coming up buildings	00	A	4
How to avoid a sprawling metropolis	00	A	1,2,3,4,5,6,7
Region sharpens attack for infrastructure funds	00	A	2,4,5,6
Managing rapid growth top priority in suburban ward	00	C	1,2,3,4,5,6
Ring-road scheme just a ruse to allow more urban sprawl	01	A	1,2,3,4,5,6,7
Now's the time to make plans for bigger city	01	A	1,2,4
Top-quality transit system must be a priority, planner says	01	A	2,4,5
Infrastructure delay shouldn't be an option	01	D	2
Biz groups getting road-weary	01	C	2,5
Growth summit important to future of our communities	01	G	1,2,3,4,5,6
Planning sustainable business parks	01	D	1,2,3,4,5,6
City drafting 'important' new blueprint for growth	01	A	1,2,3,4,5,6
Kanata died long ago, plundered by development	01	A	1,2,4,6
A blueprint for Ottawa	02	A	1,2,3,4,5,6,7

Feds to invest 6.9 M Cdn$ in west-end roads	02	D	4
City's official plan stalls on transit	02	A	1,2,4
Smart Growth policies are nothing new	02	E	1,2,3 4,5
Move ahead on the 417	02	A	4,5,6
Players pan city plan	02	C	2,4
Opportunity knocks: Report urges economic diversity	02	C	2,4,6
Economic development strategy takes shape	02	D	2,4
City's growth lies in glitz, industry	02	A	2,4,6
Ottawa Transportation Master Plan Workshop	02	C	5
Ottawa council passes official plan, prepares for legal fight with builders	03	A	2,3,6
Kanata expansion needs 128 M Cdn$ – Who will foot the bill?	03	A	2,3,6
Growth plan scores – committee backs Corel Centre-area concept	03	A	3
Ottawa readies for growth	03	F	1,2,3,4,5,6
And maybe another bridge – MPP announces ring road study	03	A	2,3,4
Mayor's frustrations build	03	B	2,4,6,7
Take a new look at this high-tech town	03	A	4
Hi-tech woes slow development plans	03	C	4
Putting Ottawa on the Tech Map	03	A	4
4 B Cdn$ building boom as good as it gets	03	A	2,4
Urban sprawl, gridlock blamed for slow rush hour traffic	03	A	1,2,3,4,5,6,7
The Queensway is big enough	03	A	1,2,4
Outdated transportation figures provide basis for new official plan	03	H	1,2,3,4,5
Our telecom sector won't be able to absorb the huge numbers of laid-off high-tech workers	03	A	4
'Cluster' economies recipes for ghost towns, Conference Board says	04	A	3,4
The worst is over – Techweekly's panel of ex-perts...Ottawa's tech scene	04	A	3,4
Widening Queensway won't help: Professor	04	A	1,2,3,4,5
Ottawa's public transit hypocrisy	04	A	1,2,3,4,5
Go west, young firms	04	C	4
West sector space glut swells	04	A	4
Ottawa's bus to nowhere	04	A	1,2,3,4,5
Grands parleurs, petits faiseurs	04	I	1,2,4,5,6
O-Train is essential to city's growth	04	A	2,5, 4,6
Route reductions run counter to city's goal of achieving 'smart growth'	04	A	1,2,4,5

*Legend for newspapers: A = Ottawa Citizen; B = Ottawa Journal; C = Ottawa Sun; D = Ottawa Business News; E = The News; F = Globe and Mail; G = Kanata Kourier; H = Centretown Buzz; I = La Rotonde. **Legend for topics: 1 = planning principles; 2 = development/growth visions, issues, initiatives; 3 = land use demand-supply issues, spatial patterns, trade-offs; 4 = high-tech industry; 5 = transportation; 6 = general reference to impacts; 7 = specific reference to the impacts of the high-tech industry on Ottawa's land use-transportation system relationship

Bearing in mind, therefore, that this is a chapter in a book and not one of the city's 400,000 Cdn$ consulting studies, we employ a keyword-based thematic approach in order to satisfy the context condition. Our research design criterion is to select articles on the basis of their discourse on the following fundamental, city-building topics:

- Planning principles;
- Development/growth visions, issues and initiatives;
- Land use demand-supply issues, spatial patterns, and trade-offs;
- High technology in all its manifestations (recall the discussion at the start of the chapter);
- Transportation in all its modes, and its various functional, structural, and user aspects, and as core entries,
- Articles containing general or specific references to the *impacts* of the high-tech industry on Ottawa's land use-transportation system relationship.

As shown by column 4 in Table 1, many of the articles cover two or more topics. Further, examination of the articles reveals that no topic 'belongs' to any group of people of influence. Those findings apply to the hundreds of articles reviewed for this study, and prompt the observation that the selected articles are, in fact, typical of the reportage in this field. In addition, two key research design findings can be derived from Table 1, which support our newspaper – based approach for examining local governments' IA record.

First, the newspaper articles provide a clear, unequivocal, 25-year message of concern about the impacts of the high-tech industry on Ottawa's land use-transportation system relationship. And, importantly, this concern is shared among all the groups of people of influence noted above – politicians, local government staff, academic experts, business leaders, newspaper people, community leaders, and ordinary citizens. That is, all groups raise questions and give opinions about how the high-tech industry could and should affect the land use-transportation system relationship, and they all make suggestions about impacts to pursue (positive) or avoid (negative).

Second, over the past 25 years the messages in newspapers to local governments from the groups of people of influence share a common theme. That is, *formal measures* – policies, plans, programmes – are and will be required to ensure that the impacts of the high-tech industry on Ottawa's land use-transportation relationship are positive rather than negative in all of their environmental, social and economic aspects. Or, in other words that might better drive home the point, these inevitable impacts are not always going to be good for Ottawa, nor are they going to be self-correcting, nor are they always going to achieve 'highest and best' status on their own. As a result, local governments need to have measures in place to minimise, mitigate, maximise, etc., the consequences associated with the impacts.

Taken in combination, the directed messages of the published articles, their large numbers, and the involvement of all the various groups of people of influence who have engaged in the public discourse on this topic, converge to provide unequivocal support for the paper's basic premise. That is, in view of the detailed and informative 'heads-up' that has been given by 30,000+ newspaper articles, it is reasonable to expect that the impacts of the high-tech industry on Ottawa's land use-transportation system relationship have been, are and will be explicitly and thoughtfully incorporated into the City of Ottawa's formal planning documents – official plans, strategic plans, development plans, transportation master plans, and other planning-related documents. This expectation is explored in the next section of the chapter.

Impact Assessment Criteria

A review of the 1970s planning documents yielded a number of 'shoulds' to be achieved or avoided, depending upon the perspective used. A selection of these 'shoulds' are presented in Table 2 as criteria that *could* be used to assess the impact of the high-tech industry on the land use-transportation relationship. That is, all the criteria selected from the 1970s documents are recently or currently employed in City of Ottawa documents on land use and transportation planning issues, goals, objectives and principles. As a result, they constitute a substantive set of measures, indicators or variables that could be used in empirical, data-based studies to assess whether Ottawa's land use-transportation relationship has been positively or negatively impacted by decisions and developments involving the high-tech industry. And, of particular import to this chapter, they can therefore be used to review local government documentation on the topic.

The general content and purpose of Table 2 is self-evident from its structure, and only a brief comment is needed to establish the difference between the two sets of criteria. On the one hand, there are positive impact criteria, which for impact assessment purposes can be expressed as measures, indicators or variables. These criteria represent outcomes to be sought as a consequence of decisions and developments that bring the high-tech industry into the land use-transportation relationship.

Conversely, negative impact criteria, which can also be expressed as measures, indicators or variables, refer to outcomes to be avoided or minimised. These outcomes include unforeseen costs, problems, disruptions or other nasty surprises that arise when things go wrong, and can be caused by errors of omission or commission or, in a worst case scenario, both types of error.

Table 2. Criteria Identified in Ottawa Area Planning Documents that *Could* Be Used to Measure and Assess the High-tech Industry's Impact on Ottawa's Land Use-Transportation Relationship

Positive Impact Assessment Criteria
• Improved safety for transit users
• Improved comfort for transit users
• Improved convenience for transit users
• Improved travel times for transit users
• Improved freedom of movement for transit users
• Increased concentration of work-trip destinations
• Increased dispersion of work-trip destinations
• Increased flexibility of transportation systems
• Increased reliability of land-use forecasts
• Increased transit service to areas other than downtown
• Increased transit ridership
• Reduced free parking downtown
• Increased concentration of denser development within the Greenbelt
• Development outside the Greenbelt concentrated in growth areas
• Reduced diversity of trip origins and destinations
• Increased decentralisation as telecommunications improve
• Increased protection of residential areas from intrusive through traffic
• Increased employment densities outside downtown
• Stricter land use controls
• Negative Impact Assessment Criteria
• Increased congestion downtown
• Increased driver-only work trips
• Reduced levels of service at arterial intersections
• Increased sprawl
• Increased costs for transit users
• Increased travel times
• Increased costs of vehicle operation
• Increased transportation infrastructure costs

Source: RMOC (1972; 1973).

In addition to emphasising the difference between positive and negative IA criteria, there are three research design features of Table 2 that need to be made explicit. First, the list of normative planning processes, outcomes and instruments used to construct the table is *indicative* of those contained in the reviewed documents (RMOC, 1972, 1973). We believe that an indicative list is appropriate and sufficient for this study, and that a comprehensiveness listing exercise is one that could be undertaken by the City of Ottawa.

Second, we employ the term *criteria* in the sense that the statements in Table 2 could be cast as variables, which could then be incorporated in analysis/synthesis techniques to measure and assess whether and how well planning goals, objectives, principles or practices that were articulated in the 1970s were achieved through the 1980s and 1990s to the present. Again, this argument is equally applicable for data-based or document-based investigations, with the only significant differences between the studies being in the particular analysis/synthesis techniques employed, and the interpretations of the qualitative, quantitative or visualisation outputs arising from the analysis/synthesis procedures.

Third, for reasons of article length and resource constraints, we could not employ all the criteria listed in Table 2 to review the impact assessment intentions or activities of local governments since the early 1970s. As a result, no attempt is made at this stage to achieve exclusivity, or to avoid collinearity, double-counting, etc., in presenting the array of *potential* variables that could be used.

The next stage in the IA process is to specify the criteria *or* variables to be used to review local governments' record of assessing the impact of the high-tech industry on Ottawa's land use-transportation relationship. Resources permitting, this activity could have been based on a Delphi exercise (see Novakowski, 1999), or it could have been done using the variable selection process designed and tested for the Walking Security Index (Wellar 1997; 1998). However, available resources did not provide the opportunity for pursuing either of those paths, so we opted for the only route open to us. That is, and bearing in mind that this is an exploratory rather than a confirmatory piece of research, we derived the variables to be used by:

- Ensuring their presence in planning documents published by Ottawa area local governments over the duration of the study period.
- Selecting the variables which, based on our knowledge of the Ottawa economy and geography, the high-tech industry, and the land use and transportation systems, appeared most likely to be included in local government assessments of the impact of the high-tech industry on Ottawa's land use-transportation relationship.

The results of our deliberations are presented in greater detail in Table 3, which contains the criteria *initially* selected for the impact assessment review. Each criterion is accompanied by an explanatory note of an if-then nature, or as a declarative statement. We hasten to add that since these brief comments are indicative in nature, no attempt is made to achieve one-to-one correspondence between every phrase in Table 3 and every planning document issued by Ottawa area local governments over the past several decades. However, on the basis of multiple readings of entire texts and re-readings of the directly pertinent sections, we submit that the criteria described in Table 3 are *overwhelmingly consistent* with the intent and content of those documents, and the opinions of those who wrote and approved them.

Table 3. Criteria from the 1972 Regional Transportation Study Initially Selected for Assessing the Impacts of the High-tech Industry on Ottawa's Land Use-Transportation Relationship

Positive Impact Criteria	Negative Impact Criteria
P1. Expansion of the transit system. If the transit system is expanded, then demand for road system infrastructure expansion is diminished as fewer cars are on roads, using parking lots, clogging intersections, obstructing transit vehicles, conflicting with pedestrians and cyclists, etc.	N1. Expansion of the regional road network via widenings, new sections, new bridges, new signals. These initiatives involve increased capital and maintenance costs, loss of land, increased environmental impacts, and invariably lead to more of the same kinds of expansion initiatives elsewhere on the network.
P2. Increased share of work trips by transit. If more work trips are made by transit, then fewer trips are made by car, and air quality is improved, crash/collision levels are reduced, law enforcement costs are reduced…	N2. Increased car-based share of work trips. Increased car-based share of trips means increased consumption of fossil fuels, increased pollution, and increased demand for more roads and widenings (see N1).
P3. Increased usage across the regional road network during peak hour traffic times. If more of the road network is used, then the traffic load at all intersections and on all road sections is diluted, and network expansions are not required.	N3. Increased congestion at intersections and/or along road segments. Increased congestion means increased time spent in vehicular travel (car, truck, bus, bicycle), increased pollution, decreased safety, comfort and convenience of all road users, and increased demands for roadway and/or intersection widenings (see N1).
P4. Better integration of land uses with high trip generation rates and high-capacity transportation services. Better connections between high transportation demand and transportation capacity means fewer mis-matches between demand and supply, and better use of resources.	N4. Increased work-trip travel times. Increased time spent in vehicles means less time for family and non-trip activities, increased demand for more expenditures on roads and/or public transit, and increased pressure to set signal lights to favour carstrucks plus passengers at the expense of transit users, cyclists and pedestrians (see N1).

P5. Better integration of transportation and industrial infrastructure. Tight(er) linkages between infrastructure components means increased efficiencies and less land used for infrastructure provision.	N5. Increased proportion of land assigned to car-related uses. More land assigned to roads, rights-of-way, driveways and parking lots means less land for residential, institutional, recreational, institutional and ecological uses/purposes, or an outward movement of the urban boundary.
P6. Reduction of critical traffic demands in centrally-located travel corridors. High-level congestion or so-called 'gridlock' in some corridors would be significantly reduced if car and/or truck demand is staggered (spread over an extended span of time), or other strategies are employed to deal with the relatively few 'choke points' which are allowed to generate network-type impacts.	N6. Localised official plan amendments that distort region-wide land use objectives. Plan amendments that are localised, that is, site-, situation-, or application-specific amendments disrupt the orderly assignment of land uses across the planning area (formerly the Region of Ottawa-Carleton, now the City of Ottawa) with attendant economic, social and/or environmental costs for the city as a whole.
P7. Per capita reduction in public expenditures annually on transportation infrastructure. Lower capital costs for roads, bridges, overpasses, transitway segments and stations, equipment, traffic signals, sidewalks, bike paths, etc., mean a reduced tax burden for individuals and businesses, or shifts in budget allocations to more valued public goods and services.	N7. Localised official plan amendments that distort region-wide transportation goals and objectives. Plan amendments that add a lane here, close a roadway there, or move buses from Street A to Street B without regard to their network consequences tend to perpetuate underlying structural or functional problems rather than provide solutions.
P8. Per capita reduction in expenditures on transportation operations. Lower expenditures for servicing and maintaining roads, transitways, buses, vehicles, paths, etc. mean a reduced tax burden for individuals and businesses, or shifts in budget allocations to more valued public goods and services.	N8. Localised infrastructure expenditures that distort region-wide development objectives. If a large share of Ottawa's budget annually goes to expanding, maintaining or reconstructing infrastructure (e.g., sewer and water) in one or two or three locations across the city, then other areas get less and may suffer sewer-related and water-related consequences that have health as well as financial implications.
P9. Per capita reduction in fossil fuel consumption for work-trip purposes. Lower consumption of fossil fuels reduces pollution, ground level ozone or smog, and contributes to improved health of individuals on or along roadways.	N9. Transportation sector expenditures that distort budget allocations. If a larger share of Ottawa's budget is spent on transportation personnel, adding to the road network, constructing sidewalks and curbs, purchasing buses, and developing a light rail system, then a smaller share is available, *ceteris paribus*, for other items.

P10. Increased number of home-based employees. Increased access to and use of telecommunications systems and services moves data and information instead of people, reduces work trips, increases trip flexibility, lowers demand for roads, reduces congestion at peak hours, reduces pollution, etc.	N10. Failure to properly maintain road network infrastructure. If Ottawa cannot properly carry out this function, then the network deteriorates, re-construction costs increase, traffic problems arise (see N3), calls for greater road financing arise (see N1, N9), and the city faces liability suits for negligence, lack of due diligence, etc.

Source: RMOC (1972).

It follows, therefore, that it is reasonable to expect that these are most likely to be among the criteria or variables included in city or regional government projects or studies assessing the impacts of the high-tech industry on Ottawa's land use-transportation relationship.

With respect to the derivation of the criteria, they represent aspects of the impact-relationship connection that are emphasised, highlighted, underlined, criticised, extolled, etc., in numerous Ottawa planning documents over the past 30 years. Further, the criteria can be translated as measures, indicators and variables that define many of the dominant issues, concerns, priorities, principles, 'druthers,' exhortations, and schools of thought which have marked the urban land use and transportation planning literature in Canada, the US, and abroad for 10, 20, 30, 40, or more years. For those reasons, as well as impressions gained from professional involvement by the authors in land use and transportation matters in the Ottawa region (beginning in the early 1970s in the case of Wellar and the 1990s for Novakowski), the criteria in Table 3 are selected as the foundation set for this study.

To summarise the contents of Table 3, the positive impact criteria refer to outcomes to be pursued or achieved in conjunction with, or as a consequence of decisions and developments that involve bringing the high-tech industry into the land use-transportation relationship. And, conversely, negative impact criteria refer to outcomes to be avoided because the costs, disruptions, afflictions, inequities, externalities, or dis-benefits that arise or accrue as a result of the high-tech industry's impacts on Ottawa's land use-transportation relationship are deemed to be unwanted, inappropriate, and unacceptable.

A full application of all the criteria in Table 3 in a robust impact assessment review would entail a study that is likely broader in scope, deeper in depth, and more sophisticated in methodology than any land use or transportation study undertaken/funded by the Regional Municipality of Ottawa-Carleton or an associated local government over the past decade. Clearly, such a study is not to be attempted here. Rather, what is appropriate for our chapter is to select several criteria to test the efficacy of the document-based IA approach, and to provide a preliminary indication of the extent to which planning documents have regard for the impact of the high-tech industry on Ottawa's land use-transportation relationship.

Document Review Process

The document review process used a keyword procedure for examining both hard copy and electronic files. This approach increases the likelihood of more quickly identifying relevant statements, while reducing the amount of time and effort required to examine documents. The document search keywords listed in Table 4 include terms from the 1970s and post-1970s local government planning documents, the newspaper articles presented in Table 1, and the learned literature in the fields of high technology, planning, transportation and impact assessment. In view of the exploratory nature of the study, the keyword upper limit was initially set at 18 and then increased to 24, with emphasis on terms that appear to best represent the theme of the chapter; that is, examining local governments' record of assessing the impact of the high-tech industry on Ottawa's land use-transportation relationship.

Table 4. Terms Used to Search City of Ottawa Official Plan Documents for Statements on Assessing the Impact of the High-tech Industry on Ottawa's Land Use/ Transportation System Relationship

capacity	ecological	industry
cause	ecosystem	integration
cluster	electronics	manufacturing
communication	effect	mitigate
computer	expansion	sprawl
congestion	gridlock	synergism
contraction	impact	tech/nology/nological
decrease	increase	tele/communication(s)

Findings from the Document Review

The focus of our examination in the document review is the City of Ottawa's most recent body of official plan documentation, and two reports which appear most relevant to our study:

1. The Official Plan (City of Ottawa, 2003a);
2. The Transportation Master Plan (City of Ottawa, 2003b).

The results of our review are summarised in Table 5. As shown by the string of consecutive *No* entries, it is our finding that not even one of the 20 impact assessment criteria is addressed in the 2003 Official Plan documents that we reviewed. This outcome is so contrary to expectations that it caused us to re-visit the Official Plan materials several times, only to obtain the same result: Table 5 is an accurate account of the total lack of regard given in the Official Plan to addressing the impacts of the high-tech industry on Ottawa's land use-transportation relationship.

Table 5. Local Government Regard for the Impact of the High-tech Industry on Ottawa's Land Use-Transportation System Relationship: Results from the Review of the 2003 Official Plan*

Potential Impacts of the High-tech Industry on Ottawa's Land Use-Transportation System Relationship: Derived from Review of Early-1970s Documentation**	**Impact Addressed in 2003 OP?**
Positive Impact Criteria	
P1. Expansion of the transit system.	No
P2. Increased share of work trips by transit.	No
P3. Increased usage across the regional road network during peak hour traffic times.	No
P4. Better integration of land uses with high trip generation rates and high-capacity transportation services.	No
P5. Better integration of transportation and industrial infrastructure.	No
P6. Reduction of critical traffic demands in centrally-located travel corridors.	No
P7. Per capita reduction in public expenditures on transportation infrastructure.	No
P8. Per capita reduction in expenditures on transportation operations.	No
P9. Per capita reduction in fossil fuel consumption for work-trip purposes.	No
P10. Increased number of home-based employees.	No
Negative Impact Criteria	
N1. Expansion of the regional road network via widenings, new sections, new bridges, new signals.	No
N2. Increased car-based share of work trips.	No
N3. Increased congestion at intersections and/or along road segments.	No
N4. Increased work-trip travel times.	No
N5. Increased proportion of land assigned to car-related uses.	No
N6. Localised official plan amendments that distort region-wide land use objectives.	No
N7. Localised official plan amendments that distort region-wide transportation objectives.	No
N8. Localised infrastructure expenditures that distort region-wide development objectives.	No
N9. Transportation sector expenditures that distort budget allocations.	No
N10. Failure to properly maintain road network infrastructure	No

* Reviewed documents include the 2003 Official Plan for the City of Ottawa and the Transportation Master Plan that supports the Official Plan (City of Ottawa, 2003a, 2003b). ** The potential impacts are from Table 3.

In attempting to reconcile expectations and findings, or to understand the reasons for their differences, it is necessary to begin by questioning the study's expectations and research design. They may have been unrealistic or inappropriate to begin with, thereby creating a situation in which it is unlikely or impossible for the findings to 'measure up.' By way of illustration, if the chapter's research design involves a matter that is outside the purview of the Official Plan, then it is not reasonable to expect the (impact assessment) matter to be addressed in Plan documentation. As the next several paragraphs demonstrate, there is every reason to believe in the validity of both the study's expectations and its research design.

First, and based on the 1970s planning documents which made an eloquent case more than 30 years ago about the importance of assessing impacts, and the many thousands of associated media items published over the past 30 years (see Table 1 for a selection of indicative articles), it is reasonable to expect that the review of 2003 Official Plan documents would yield numerous directed, substantive statements addressing both the desired impacts and the to-be-avoided impacts of the high-tech industry on Ottawa's land use-transportation system relationship. Indeed, we anticipated encountering so many pertinent statements that we explicitly noted in the heading of Table 3 that the assessment criteria represented an *initial selection*. Following that design decision, we then intended to restrict the number of criteria actually used in the official document review in order to place an upper limit on the size of the chapter.

Second, in support of our intention to restrict the number of criteria used in the official documents review, our expectation of finding numerous references on impact-related matters in the 2003 Official Plan was tangibly and virtually reinforced over the course of the study. That is, we were accumulating a large and growing collection of newspaper articles, news releases, and website postings on the City of Ottawa's high-tech future, its smart growth summits and conferences, its smart city connections, and its various transportation futures with and without the O-train, more bridges, ring roads, an enlarged Queensway, more dedicated bus lanes, and so on. Moreover, it warrants emphasising that these materials did not represent just the interests of politicians, citizens, academics, business people and consultants: they also represented the opinions of planners themselves.

It was a reasonable expectation, therefore, beginning with the 1970s planning documents, and continuing through the 1980s and 1990s with a steady flow of local government projects and reports on the high-tech industry, that Ottawa's 2003 Official Plan would contain a series of prominent statements explicitly dealing with the impacts of the high-tech industry on Ottawa's land use-transportation system relationship. Indeed, based on the volume of newspaper items on this topic week after week for more than 20 years, and the full participation of all groups of people of influence in the public discourse, it was reasonable to expect the 2003 Official Plan to address each impact assessment criterion presented in Table 5.

However, on the basis of what we actually encountered in our review of the 2003 Official Plan, our expectations were sorely misplaced. Instead of finding the rich body of documentation that is needed to justify the claim of being a 'world class' city, or even a 'well-planned' city, what we discovered is that the City of Ottawa is totally silent on a matter which, on paper, has been of deep and abiding concern to politicians, residents, academics, business people, journalists and planners for 30 years!

As for the validity of the findings of the document review presented in Table 5, we submit that they represent reality to a high degree of accuracy. That is, insofar as we are able to ascertain, the City of Ottawa's 2003 Official Plan does not use any of the impact assessment criteria in Table 5, nor any other impact assessment criteria which we could identify, in seeking to ensure that Ottawa's land use – transportation relationship is positively affected by the high-tech industry. That being the case, it is logically appropriate to contemplate the following question:

> Why is there a major difference between what was expected and what was found regarding the extent to which the Official Plan addresses concerns about the impact of the high-tech industry on Ottawa's land use-transportation relationship?

In terms of explaining the findings contained in Table 5, we were not retained by the City of Ottawa to advise or consult on any aspect of the 2003 official plan process. Consequently, we have no insiders' knowledge of how it came to pass that regard for the impact of the high-tech industry on Ottawa's land use-transportation system relationship is totally excluded from the 2003 Official Plan. Further, despite many years of involvement in the planning field, numerous local government consulting engagements, and active participation in dozens of municipal and regional official plan reviews, we have no outsiders' explanation for what appears to be a gross error of omission and/or commission by the City of Ottawa in adopting an official plan that makes no mention of the impact of the high-tech industry on the land use-transportation relationship.

Conclusions

The comments in the last section summarise our findings with regard to the chapter's first research objective, which is to ascertain the extent and degree to which local governments in Ottawa-Carleton and later the City of Ottawa had and have regard for the impact of the high-tech industry on Ottawa's land use transportation system relationship. That is, the chapter began by presenting a selection of the impact assessment-related statements of the early 1970s, and then it presented dozens of newspaper items selected to represent the many thousands of editorials, columns, letters, reports, etc., that typify the public discourse on impact-related topics over the past 20+ years. However, and despite high expectations that it would be otherwise, Table 5 reveals that the 2003 edition of the City of Ottawa's Official Plan is silent in terms of its regard for the impact of the high-tech industry on Ottawa's land use-transportation relationship.

As for when and why that shift occurred, and how it came to pass that a topic of deep and broad public discourse is excluded from the Official Plan, those are questions which are beyond the purview of this current investigation. What can be said, however, is that we have no research-based explanation for the difference between what we reasonably expected of the 2003 Official Plan, and what was actually found after intensive keyword-based document searches, and three thorough, page-by-page readings of the documents.

As for the second research objective of employing a document-based approach for impact assessment review purposes, the successful completion of the chapter is evidence that historic documents can provide a substantive basis for such a study. In this investigation, we were able to identify several 'foundation' documents, develop a framework for tracking a 30-year record and, in the process, make a substantial contribution to IA methodology. That is, we show how both positive and negative impacts can be explicitly incorporated in a local government performance review. And, we also show how a text-based comparative study approach can reduce many hundreds of pages of text to the several critical pages, tables or paragraphs that reveal the extent to which impact assessment methodology is (or is not) employed by local governments in the development and implementation of policy, planning or programme objectives.

References

City of Ottawa. (2003a). Ottawa 20/20: official plan. *City of Ottawa*. Retrieved December 15, 2005 from http://ottawa.ca/cityservices/planningzoning/2020/op/index_en.shtml.

City of Ottawa. (2003b). Ottawa 20/20: transportation master plan. City of Ottawa. Retrieved December 15, 2005 from http://ottawa.ca/city_services/planningzoning/2020/transpo/ toc_ en.shtml.

Harrison, R., Cooper, S. and Mason, C. (2004). Entrepreneurial activity and the dynamics of technology-based cluster development: the case of Ottawa. *Urban Studies, 41* (5/6), 1045-1070.

Novakowski, Nick. (1999). *Derivation of an evaluation instrument for judging the quality of ecosystem-based municipal plans*. PhD diss., University of Ottawa.

Ottawa Business Journal. (May 14, 2004). Battle for Ontario moves into tech sector. *Ottawa Business Journal*, May 14, 2004.

RMOC (Regional Municipality of Ottawa-Carleton). (1972). *Progress report: transportation study*. Ottawa: Regional Municipality of Ottawa-Carleton.

RMOC (Regional Municipality of Ottawa-Carleton). (1973). *Travel forecasts and analysis*. Ottawa: Regional Municipality of Ottawa-Carleton.

Smith, R., Wellar, B. (1992). A progress report on public policy objectives achieved through IS/GIS/LIS… In *IS/GIS/LIS and Policies, Plans and Programs: Thirty years in Perspective*. Vol. 5, 1992 URISA Proceedings. Washington, DC: Urban and Regional Information Systems Association.

Statistics Canada. (2004a). Population and dwelling counts, for Canada, census metropolitan areas, census agglomerations and census subdivisions. *Statistics Canada*. Retrieved on December 23, 2005 from http://www12.statcan.ca/english/census01/products/standard/popdwell/Table-CMA-C.cfm?CMA=505).

Statistics Canada. (2004b). Community profiles: 2001. *Statistics Canada*. Retrieved on December 23, 2005 from http://www12.statcan.ca/english/Profil/ PlaceSearchForm1.cfm.

Statistics Canada. (2004c). Community profiles: 1996. *Statistics Canada*. Retrieved on December 23, 2005 from http://www12.statcan.ca/english/Profil/PlaceSearchForm1.cfm.

Statistics Canada. (2004d). *Level of educational attainment*. Catalogue No.97F0024XIE2001012. Ottawa: Statistics Canada.

Statistics Canada. (2004e). *Federal government employment, wages and salaries, by census metropolitan areas*. CANSIM Table 183-0003. Retrieved on December 23, 2005 from http://www.statcan.ca/english/Pgdb/govt58a.htm.

Statistics Canada. (2006). *Preliminary 2006 CMA: Ottawa-Gatineau, Ont./Que*. Catalogue No.92F-0138-MIE-2003-002. Ottawa: Statistics Canada.

Wellar, B. (1975). *Housing for the future: questions and arenas for action*. B.75.8. Discussion paper. Ottawa: Ministry of State for Urban Affairs.

Wellar, B. (1981). Economic development strategies: a Canadian context. In *National and regional economic development strategies: perspectives on Canada's problems and prospects*. Edited by Barry Wellar. Ottawa: University of Ottawa Press.

Wellar, B. (1990). Electeds, planners, people, and official plans: their new roles in defining and directing regional development. In *Focus for the Future Conference and Workshop Proceedings*. Waterloo, ON: Regional Municipality of Waterloo, Planning and Development Department.

Wellar, B. (1996). The fundamental importance of local economies. *New City Magazine*, *12*, 58-63.

Wellar, B. (1998a). Combining client-driven and curiosity-driven research in graduate programs in geography: Some lessons learned and suggestions for making connections. *Papers and Proceedings of the Applied Geography Conferences*, *21*, 213-220.

Wellar, B. (1998b). *Walking security index*. Ottawa: University of Ottawa.

Wellar, B. (2000). *Newspapers as a source of fact and opinion on pedestrians' safety, comfort, convenience: a keyword-based literature search and review*. Ottawa, ON: City of Ottawa and University of Ottawa.

Wellar, B. (2003). Connecting the national urban strategy, planning, and the urban sustainability index. *Plan Canada*, *43*, (4), 42-44.

Wellar, B. (2004). *Fixing the Ontario municipal board: a strategic approach for citizen groups*. Federation of Canadian Neighbourhoods. Retrieved January 2, 2006 from www.urbanneighbourhoods.ca/Info/fixOMB.doc.

Wellar, B., Novakowski, N. (2000). Detailing the implications of ecosystem-based planning for applied geography training and research project designs. In *Papers and Proceedings of the Applied Geography Conferences*, *23*, 215-223.

Wolfe, D. (2002). *Knowledge, learning, and social capital in Ontario's ICT clusters*. Retrieved January 2, 2006 from http://www.utoronto.ca/progris/web_files/recentpub.htm.

CHAPTER 4

Unanticipated Benefits

The Role of Planning in the Development of the Ottawa Region Technology Industries

David GORDON and Betsy DONALD,
with John KOZUSKANICH

School of Urban and Regional Planning, Queen's University
Department of Geography, Queen's University
Department of Civil Engineering, Queen's University

Introduction

Ottawa was not an attractive place at the beginning of the 20th century, being closer to a frontier lumber town than a capital city. William Lyon Mackenzie King arrived at a particularly bad time in July 1900, shortly after a lumber fire had destroyed 14% of Ottawa and half of Hull:

> ...the first glimpse of the city was from the lately fire swept district and it was gloomy enough. The business part of the town is small and like that of a provincial town, not interesting but tiresome. I will miss greatly the University society and the pleasant surroundings of Cambridge. Ottawa is not a pretty place save about the Parliament Buildings and has all the non-attractions of a small town. I find myself bearing up against a load on this place (King, 1900).

King's observations are interesting, not just because he later became Canada's longest serving Prime Minister (1921-30; 1935-48). King was a highly-educated young researcher brought to Ottawa from the midst of his Harvard doctoral studies in economics. Within a year, he was the capital's youngest deputy minister and a protégé of the Governor General, yet King despised the city. As Prime Minister, he pushed hard to re-plan Ottawa to make it an attractive place for diplomats, politicians, civil servants and visitors. But King had limited success, since the effects of the Depression and two world wars left Ottawa dirty, overcrowded and worn down in 1948. King would probably have been pleased by the appearance of Canada's capital city in 2006, but he would have been surprised to hear confident predictions that the federal government would be surpassed as the region's largest employer by high-technology industries that he could not have imagined fifty years earlier.

In this chapter, we argue that the planning and development that transformed the Ottawa-Gatineau region from a lumber town into a modern and attractive national capital also had the unanticipated benefit of creating good conditions for growing high-technology industries. Doyle (1991) named this transformation "from white pine to red tape to blue chips." We begin with a review of the literature relevant to Ottawa's high-tech cluster and a description of the federal government's seven decades worth of efforts to redevelop the capital city. We then consider the effect of local and regional planning from 1960-90 and the micro-geography of the high-technology cluster. We conclude with discussion of regional planning and examine development policies for high-technology industries in the Ottawa-Gatineau area.

Theoretical Background and Literature Review

For decades, urban planners have been taking quality of life indexes into account when planning for the long-term health and sustainability of their cities and regions (Dissart and Deller, 2000). However, it is only in the last few years that planners have considered 'quality of life' to be an essential component in their economic plan and policy-making. There are two main reasons for this heightened attention to the role that 'quality of life' now plays in urban economic development. The first concerns the importance of *innovation* in initiating and sustaining a robust knowledge-intensive regional economy.

Innovation in this context includes the ability of firms and people in the region to create new ideas in terms of new products, processes, designs and marketing (Best, 2001). Furthermore, the literature suggests an important *social* dimension to this innovation (Granovetter, 1985; Dosi, 1988). In other words, innovation is not something that only happens when a scientist works in isolation in an R&D lab, but it is believed to increasingly occur when people work together in teams in the same place (Gertler, 2001; Lundvall, 1998; Porter, 1999). Sometimes this innovation is a result of codified knowledge transfers between groups of knowledge-workers either working within the same firm or between firms. Other times it is a result of more tacit knowledge transfer either between people or between firms.

Knowledge-transfer can occur spontaneously. Yet, a growing body of evidence now signals to the wider social and institutional context in either facilitating or hindering effective knowledge-transfer. In other words, the policy context matters. Regional and extra-regional institutions can facilitate the transfer of knowledge across space. These institutions include institutions of higher education and technology-transfer, civic organisation, and business associations. It is this research finding that has lead local economic development planners to focus their efforts on, among things, industrial cluster formation and policy.

The second related area of focus for economic development planners has been on talent attraction and retention. If indeed innovation is now a key dimension of the knowledge economy, then the worker who generates the ideas, creativity and imagination for knowledge-intensive innovation to occur must be present. More and more research now demonstrates that talent is the critical factor of production in the new economy (Feser, 2003; Florida, 2002, 2005; Glaeser, 2000; Markusen, 2004). This research also supports the hypothesis that this skilled worker is also highly mobile and attracted to places that offer certain employment opportunities and 'quality of life' amenities and lifestyle options.

Richard Florida with his best-selling book, *The Rise of the Creative Class* (2002) and more recently *The Flight of the Creative Class* (2005), has perhaps been the most eloquent and influential advocate for linking quality of life amenities with talent attraction and retention. To compete in the age of talent, argues Florida, cities and regions must promote 'quality of life' and amenity characteristics that appeal to talented and creative workers. Florida's research found certain aspects of quality of life to be particularly appealing to knowledge-workers, or more specifically, to his study group of young, high-technology talent. First, these knowledge-workers want to live in places with a *thick labour market*. Given the volatility of the high-tech sector and the frequency with which people change jobs, the places that can offer a variety of good job opportunities will be favoured over others. The second major finding of his research was that knowledge-workers in technology-intensive sectors want to live in places with well-developed amenities, particularly in the form of nature, recreation, lifestyle, and arts and entertainment.

The third major finding was that young knowledge-workers in particular want to live in places that had a culture characterised by vibrancy, diversity and tolerance. Florida views cultural vibrancy as a product of a series of amenities mainly around a hip and vibrant street culture, including outdoor cafes, trendy restaurants, local art galleries and a lively music scene (Florida and Gates, 2001). Tolerance and diversity, on the other hand, include a tolerance for a diverse range of lifestyles and sexual orientations.

What then, are the implications of this research for local economic development planning? According to Florida, these 'quality of life' factors must be central elements in a city's strategy to attract knowledge-workers and build innovative economies. It is not about competing for business attraction with cheap labour and low taxes. Rather, the key to a region's prosperity and success is shaped much more by attracting talent, and this is done through quality of life amenity strategies.

From a policy and planning perspective, these quality of life amenities strategies include policies towards culture and the arts, land-use planning, immigration and settlement, education as well as the role that political culture and local economic development policy play in shaping the health and future of cities. More specific elements of quality of life amenity strategies include good public transit, fine public schools, universities and colleges, parks and places for jogging and cycling, plus a diverse choice of culture.

It is on this latter point of 'culture' that Florida's work has been most widely cited (Donald and Morrow, 2003). His argument has been that the policy implications that derive from notions of, and policies towards, culture and the arts are not how we have typically thought about culture. Much of the explicit public funding in the area of culture has traditionally been directed toward 'capital-C' culture and arts institutions like museums, galleries, symphony music, opera and ballet, and broadcasting – types of culture that the City of Ottawa has in abundance. Yet, this type of policy does not necessarily directly address all the types of culture that Florida is referring to. He believes planners must focus their energies on creating the conditions for vibrant and interesting streetscapes and city life, with funky cafes and restaurants, thriving music, art, literature, and design and fashion scenes. Echoing Jane Jacobs, many of the policy options centre on the use of city space along the lines of new urbanism planning. The second aspect of culture has to do with tolerance and diversity. These include policies that have supported immigration and settlement, human rights, education and citizenship education.

There is no doubt that Florida's creativity and talent policies are having a profound impact on current economic development planning across North America. At the same time, the research and its implications are facing a host of criticisms from prominent scholars (Glaeser, 2004; Peck, 2005). The criticisms are far-reaching and have to do with his research methods as well as the fast-translation of his models into public policies. The latter criticisms are relevant for this study.

In terms of fast-translation of policy applications, cities across North America, Europe and Asia have witnessed a rapid adoption of Florida's ideas into economic development practice. Two examples are Michigan's Cool Cities program and Creative Tampa Bay that call their communities to focus their energies on festivals, events, cultural activities and leveraging public and private funds to revitalise urban historic distributions through the construction of lofts and bike paths but most importantly to make sure they foster a 'quality of place' environment attractive to young, high-technology workers.

Indeed, city after city across North America has made 'quality of place' a central element of their economic development planning. However, much of this practical work has tended to conflate 'quality of life' – an individualised concept – with 'quality of place,' a concept that suggests a general consensus among actors. This assumption of a consensus tends to favour certain inhabitants over others (Donald, 2001). In particular, it tends to acknowledge and reward a very narrow definition of creativity (Peck, 2005). One of the risks of this new adoption of Florida's quality of place strategies is to simply focus on place marketing the consumptive values of funky urban districts. However, this surface level place marketing has the potential to gloss over the essential long-term investments required to maintain and enhance a creative, knowledge city.

The need to turn attention to the essential long-term planning of investments required to improve the quality of life of a city is not new to planners. It is also something that becomes very obvious when we turn to our case study of Ottawa. As we will see, Ottawa is an excellent example of how steady, balanced, long-term planning has both intended as well as unintended consequences in regional economic development. While Ottawa may not be the dynamic urban hipster-town so often celebrated in the high-tech knowledge economy literature, it has a surprisingly strong high-technology sector that has managed to maintain and attract a talented set of workers to its industries. What we discover through an examination of the history of Ottawa's planning and development is that the implementation of Ottawa's *Plan for the National Capital* and other policies during the post-war period had the unintended consequence of creating the right conditions for a high quality of life for its residents. This quality of life included – contrary to much of the current literature in this area – a relatively weak urban-hipster component, but a solid supply of 'Capital C'culture, vast swaths of green parkland, and low-density, car-oriented options for high-technology Industries and their talented employees (O'Neill, 2005).

From "White Pine to Red Tape" Planning a National Capital

The half-burned lumber town that appalled Mackenzie King in 1900 was also a concern to Canada's Prime Minister and Governor General. Sir Wilfrid Laurier's government established the Ottawa Improvement Commission (OIC) in 1899 to create a more attractive capital city. One of their earliest actions was to retain Montréal landscape architect Frederick Todd to prepare a preliminary plan for parks and open spaces in the region (Todd, 1903; Gordon, 2001a; 2002a).

The OIC began the creation of parkways along the Rideau Canal, but there was some discontent about the overall planning for the nation capital, especially in comparison to the ambitious 1903 plan for Washington and the 1912 competition for the design of Canberra in Australia. A new Conservative government under Sir Robert Borden established a Federal Plan Commission to prepare a scheme for both Ottawa and Hull. Their 1915 report was prepared by Chicago architect Edward Bennett, co-author with Daniel Burnham of the 1904 San Francisco and 1908 Chicago plans. It was Canada's first comprehensive plan for a city-region (FPC, 1915). Although the report addressed railways, streetcars, roads, services and parks, it was Bennett's City Beautiful style designs for the core of Ottawa which attracted the most attention. But the schemes to create a grand capital were put on hold during World War I and the later efforts to rebuild the Centre Block of the Parliament Buildings, which had burned in a spectacular fire in 1915 (Gordon, 1998).

The City of Ottawa began its formal planning activity with the 1921 establishment of the Ottawa Town Planning Commission under the leadership of engineering consultant Noulan Cauchon. However, City Council rarely supported Cauchon's schemes, and most inter-war planning initiatives remained with the federal government. The City co-operated with Prime Minister Mackenzie King's plans to redevelop a piece of the downtown into the National War Memorial and Confederation Square, designed by leading French urbanist Jacques Gréber (Gordon and Osborne, 2004). Most other planning and development activity ground to a halt during the 1930s Depression. According to Eggleston (1978), one of the few new projects completed during the decade were the classically designed central laboratories of the National Research Council (NRC) on Sussex Drive (Figure 1). World War II increased the pressure for change in the Ottawa region. The NRC had outgrown its Sussex Drive site before the war, and moved quickly to establish a new complex on a farm just past the eastern edge of the city. The Montreal Road laboratories were built in a low-density, campus-style, with inexpensive Modernist buildings set among green lawns and parking lots.

Figure 1. The National Research Council on Sussex Drive

Source: Gréber, 1950, Figure 67.

Other government laboratories were scattered about the city in former industrial buildings and the Defence Department built testing facilities at its rifle ranges on rural lands west of Ottawa. The war effort placed a severe strain on Ottawa's building stock and infrastructure. New public servants worked in 'temporary' wooden office buildings hastily constructed in open spaces on Parliament Hill and downtown. Little additional housing could be built and gasoline was rationed, so many civil servants boarded in existing homes and commuted to work on thirty year-old streetcars (Taylor, 1986).

As the veterans flooded back after the war, Ottawa looked dirty, old, crowded and a bit worn out, rather than the capital city of a victorious nation. Many of the industrial sites remained within view of Parliament Hill, and a great lumber fire burned both banks of the river in 1947, damaging one of the three road bridges. Frustrated by a half century of inaction, Prime Minister Mackenzie King announced that the national memorial to World War II would be to construct a capital city worthy of

Canada, rather than just another statue. King borrowed Jacques Gréber to lead the team that would prepare the National Capital Plan. The personal attention of the Prime Minister assured that the plan would have the full resources of the federal government behind it (Gordon, 2001b).

The 1950 *Plan for the National Capital* is one of the most significant documents in Canadian planning history. It was a guide for the rapid transformation of Ottawa and Hull from dreary industrial towns to an attractive, modern and functional capital (Gréber, 1950). The pattern of industrial development changed dramatically from 1950 to 1970 (Figure 2).

Figure 2. Montreal Road Laboratory of the National Research Council

Source: Gréber, 1950, Figure 66.

The new industrial parks were popular among local governments because they brought in commercial property taxes with little need for expenditure on services. Although several projects were privately owned, the NCC, Ottawa, Nepean and March townships all developed modern industrial parks to ensure that a good supply of serviced land was available for businesses. Construction of infrastructure to implement the plan began before the final document was approved in 1950 with the east-west railway line relocated south of the built up area. An expressway (the Queensway) was built by the national, provincial and local governments in the vacated rail corridor, improving accessibility to undeveloped land to the east and west. The NCC built other parkways along the Ottawa River, facilitating rapid vehicular traffic throughout the region. New suburban industrial parks replaced the crowded plots next to the rail yards, allowing more room for expansion and removing conflicts between industrial and residential uses (Figure 3).

Figure 3. 1950 Plan for Canada's Capital

Source: Gréber, 1950, Figure 128.

The plan called for a decentralisation of activities from the city core, partly for fear of nuclear attack (CPAC, 1952), but mainly to reduce congestion and open up popular new lifestyles. Federal offices and laboratories that did not require frequent contact with the public or politicians were accommodated in three spacious office parks on the edge of the countryside – Tunney's Pasture, Confederation Heights and the NRC area. Adjacent lands were designed for private development of new neighbourhoods, designed with local parks, schools and shopping centres. The new federally-insured, long-term mortgages ensured that veterans and middle class families could purchase a new house in a green suburb located only a few minutes drive from work (Harris, 2004).

The 1950 *Plan for the National Capital* proposed a greenbelt to surround the new suburbs, with satellite communities to absorb future growth, similar to the 1946 Greater London Plan. While the greenbelt was to provide visual relief and some recreational amenities to the suburban residents, it was also intended as a site for the airport and research facilities that required large tracts of land and would benefit from a non-urban location (FDC, 1958). The greenbelt became home to the Defence Research Board, the Communications Research Centre, the Agricultural Research Centre, the Defence Proving Grounds and other federal facilities. Most importantly, the Northern Electric Company (later named Bell Northern Research, Northern Telecom and Nortel Networks) was sold a 28-hectare site in the western Greenbelt to develop its research laboratories. Nortel later leased an adjacent 121 hectares from the NCC and its research campus has been the region's largest private employer and source of dozens of spin-off high-technology companies.

When Nepean and Gloucester Townships resisted the greenbelt designation, the federal government established the National Capital Commission, in 1959, with expanded powers to expropriate land and build parks and infrastructure. The 1950 *Plan for the National Capital* was probably the most thoroughly implemented Canadian plan of the 20th century. By 1970, when most of the plan had been completed, the NCC had spent 243 million (or 1.9 billion in 2004 Cdn$) to acquire land and build parks, roads, utilities and other infrastructure (NCC, 1970; Gordon, 2002b). The Ottawa region was transformed into an attractive, somewhat dull, but functional metropolis with a high quality of life for its residents. The national capital became a tourist destination and could attract and hold highly educated workers for the federal government and its associated laboratories and research facilities. It was not a cosmopolitan metropolis, but it had outstanding recreational facilities such as parkways, bike paths, the Rideau Canal and Gatineau Park just outside the doors of many residents. Many families also took advantage of high quality landscapes within an hour's drive in three directions to build cottages or ski chalets.

Local and Regional Planning: 1960-1990

Local governments had minimal roles in planning the redevelopment of the Canadian national capital region from 1945 to 1970. The City of Ottawa was busy with building a new high-rise downtown, extending services to annexed suburban land and urban renewal in the 1960s (Taylor, 1986). Its first regulating plans were adapted from the roads and parks schedules of the federal 1950 plan (Ottawa, 1953). During the 1970s, neighbourhood preservation became more important and the local government began to champion public participation in opposition to the plans of the National Capital Commission (Taylor, 1989).

The rural townships had their hands full managing the transition from farms to modern suburbs. Most local governments were scrambling to accommodate rapid growth caused by the post-war baby boom, and expansion of the federal government. Although the 1950 *National Capital Plan* projected for the region to double in size from 250,000 to 500,000 people within the new Greenbelt, this population target was reached in eighteen years rather than the 40-50 years forecast in the plan (Gréber, 1950; Ottawa, 2004). The local governments could not keep up with the demand for serviced land despite the financial assistance of the NCC for sewage treatment plants, bridges, roads, and major pipes. The land inside the Greenbelt was almost filled by the mid 1960s, and private developers began to purchase farms near the planned extensions of the Queensway, first in the west in March Township and then in the east in Gloucester and Cumberland townships.

Forward planning was minimal and most Ontario townships did not adopt official plans until the late 1960s (CPAC, 1966; 1972). The local governments in the Québec portion of the region followed in the 1970s (Ploegaerts, 1973; 1976). The NCC and local government planners were so busy accommodating the growth of the federal government that few people noticed the steady growth of the high-technology industries. As Porter (1998) has noted, the scattering of high technology across several Standard Industrial Classification codes used in the census makes it difficult to discern the emergence of a new cluster.

The smaller local governments did not have an adequate property tax base to support the infrastructure costs of rapid suburban residential development. Industrial development helped reduce the fiscal impacts of growth, so most local governments planned for large industrial parks (Elliot, 1991). Ottawa, Nepean, Gloucester and March townships developed competing facilities and the City of Ottawa formed the Commercial and Industrial Development Corporation (CIDC) in 1962 to attract new businesses (Leibovitz, 2000, p. 59). But most of the local governments were focused upon traditional manufacturing and commercial tenants; as noted by Nepean's chief planner: "No one knew we would be the high-tech capital, we were just trying to obtain the same types of regular industry as everyone else…" (Jacobs, 2004).

The exception to this trend was Kanata, a satellite community established west of the Greenbelt in the 1960s. Local builder William Teron purchased 1,295 hectares of land just outside the Greenbelt to build a 'new town' in 1959 (Kanata, 1999). Kanata was planned with high quality design values (Timusk, 1976). The first neighbourhood, named Beaverbrook, proved popular with employees of Bell Northern Research and Microsystems International, whose labs were just across the municipal border in adjacent Nepean. The new suburban residents soon outnumbered the farmers and were voted onto the council of March Township. A BNR employee, John Mlack, was given a two-year leave of absence by the firm to become the first non-rural reeve (leader of council) in 1968 (Smart, 2004).

William Teron and the new March council targeted Kanata as the technology community from the beginning. Teron named his first industrial park 'Technology Park' and aggressively pursued Atomic Energy of Canada Limited (AECL) as the lead tenant, with the direct assistance of March Township (Smart, 2004). The local government was soon targeting other technology industries for its own industrial parks (MacKay, 2004).

Kanata initially appealed to high-technology employees because of its proximity to BNR and CRC, high neighbourhood design standards and excellent recreational opportunities at the doorstep. But Teron was undercapitalised and could not afford the infrastructure investments needed for a greenbelt new town. Ottawa developer Robert Campeau took over the project in 1970, but the residents and council pushed him to keep the high community planning and development standards. Mayor Marianne Wilkinson explained: "It is imperative to maintain the quality of life... That's why the high-tech people come here and that's why they stay" (Thomas, 1983, p. 60). A new City of Kanata was carved out of rural March and Goulbourn townships in 1978 with Wilkinson as its first mayor, continuing her focus on the high-technology industries.

The social networks of the Kanata residential pioneers were an important factor in the expansion of high-technology employment of the town. Denzil Doyle, president of Computing Devices was one of the first residents in 1965. When his firm needed expansion space after being purchased by Digital Equipment Canada, he built in Kanata (Thomas, 1983, p. 61).

Terence Mathews and Michael Cowpland bought homes in Kanata when they were recruited from Britain to join Bell Northern Research. They left in 1973, forming Mitel (MIke and TErry's Lawnmowers) in Mathews' garage. By the end of the year, they had switched products to telephones and were renting space in Kanata's Junior Chamber of Commerce building. Three years later, Mitel relocated hundreds of employees to a nearby industrial park before the local services were installed (Thomas, 1986, p. 29-46). Mitel became competition to Northern Telecom before it was sold in the 1980s, making both Cowpland and Mathews extremely wealthy. Cowpland then initiated Corel, and Mathews founded Newbridge Networks. Mathews became the lead 'angel investor' in dozens of local high-technology start-up companies. Many of these high-tech start-up firms established their first offices in Kanata and adjacent Nepean, which had repositioned its industrial parks as 'business parks' to become a competitor – Figure 4 (Elliot, 1991; Doyletech, 2003; 2002).

Figure 4. Industry Relocation in Ottawa, 1948-1974

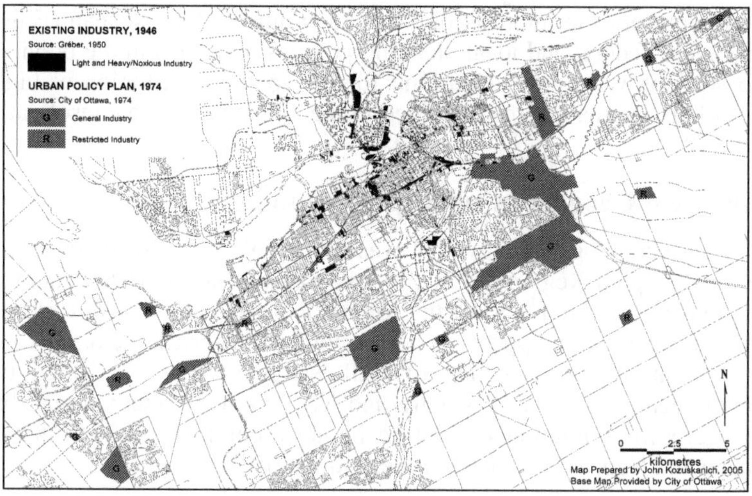

Source: Gréber, 1950; RMOC, 1974 (drawn by John Kozuskanich).

The regional infrastructure funding and planning problems were partially solved by local government re-organisations enacted by the provinces of Ontario in 1968 and Québec in 1973. The Regional Municipality of Ottawa-Carleton (RMOC) took over responsibility for regional planning and development of major roads, transit, trunk services and water and sewage treatment plants. RMOC had borrowing powers and financial services for rapid suburban development inside and outside the Greenbelt, loosely guided by the 1974 regional official plan (Ottawa-Carleton, 1974). The *Communauté régionale de l'Outaouais* (CRO) performed similar functions on the north side of the Ottawa River (Outaouais, 1976).

Most of the local governments remained in place from 1970-2000, with reduced responsibilities for local planning, economic development and local services. The local plans concentrated in generic land use planning issues and none addressed the growth of high-technology industries. Our review of 54 plans from 17 different local and regional agencies from the period 1957-1996 found only one reference to high technology – in the 1990 revisions to the Kanata Official Plan. Most plans addressed employment as a generic land use issue and none of the local or regional plans had any reference to the federal government, the region's largest employer.

The municipalities continued to compete for industrial development, but the regional governments gradually took the lead in this area. The CIDC was renamed the Ottawa-Carleton Economic Development Corporation in 1985, although there was continued grumbling that it did not fairly represent the interests of local governments like Nepean (Elliot, 1991). The federal and local planning policies described in the previous two sections were the background for the spatial distribution of high-technology companies that emerged in the 1980s and were consolidated in the 1990s.

The Local Geography of High Technology

High-technology employment was not evenly distributed across the National Capital Region (NCR). It was spatially concentrated south of the Ottawa River and west of the Rideau River (Figure 5). Most of this spatial concentration appears to be the work of historical factors such as the original western location of Bell Northern Research/Nortel, the Communications Research Centre and Microsystems International, but the aggressive targeting of high-technology employees and firms by Kanata and Nepean may have influenced the pattern.

Figure 5. Ottawa Region Business Park Location: 2004

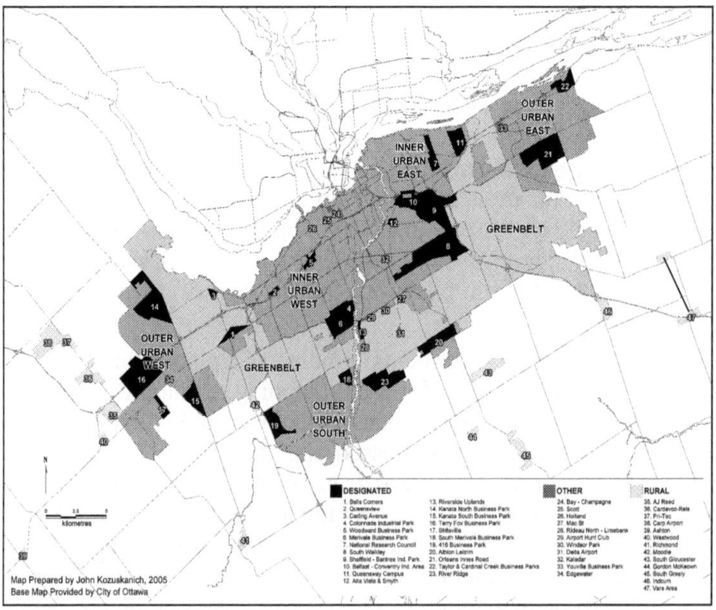

Source: Drawn by John Kozuskanich.

For example, the National Research Council is located in the east end of Ottawa, and about half of its start-up companies had their initial premises in the east end (Figure 6). But, some of these firms may have drifted westward when they expanded, since the overall pattern of employment in the late 1990s is strongly skewed to the southwest of the NCR for all high-tech sectors except software consulting, which is concentrated near the Ottawa downtown (Figure 7).

Figure 6. Ottawa Region High-tech Employment: 2004

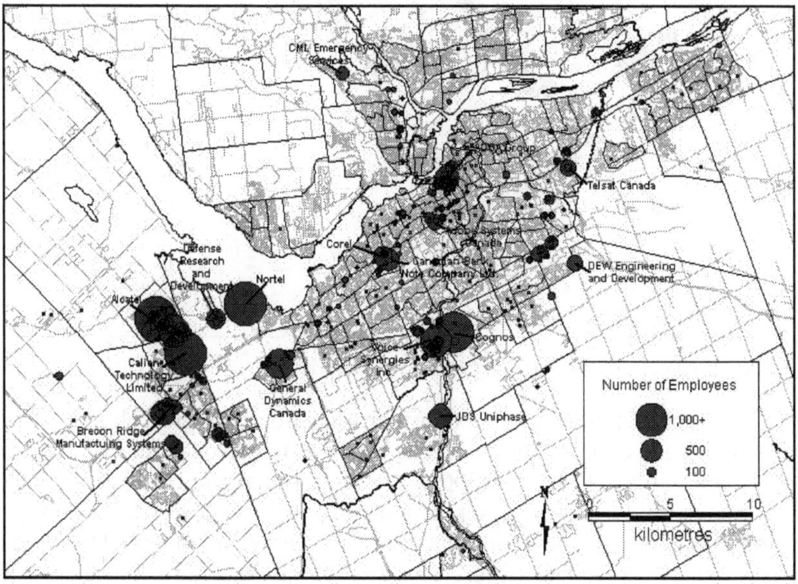

Source: Ottawa, 2004; OCRI, 2004; Drawn by John Kozuskanich.

Figure 7. Location of Spin-off Firms from the National Research Council

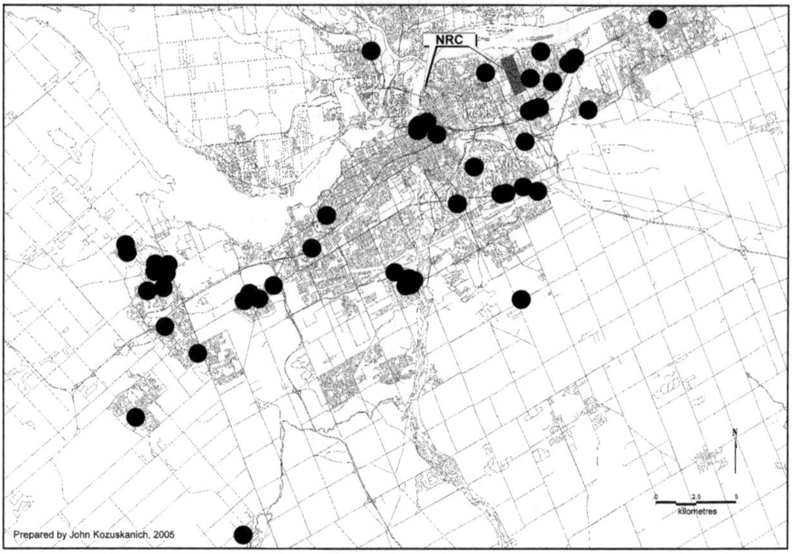

Source: Doyletech, 2002; Drawn by John Kozuskanich.

The factors that influence the micro-location of high-technology industries were explored in a recent study conducted by the Corporate Research Group (Nixey, 2000). The firm conducted structured key informant interviews with senior executives of the eight largest high-technology firms. These organisations accounted for over 20,000 jobs distributed over 100 facilities. This was complemented by a telephone survey of 100 of the 470 small firms (defined as 1-100 employees).

For the large companies, the most important site selection criterion was a large supply of inexpensive parking. Other important dimensions were expansion potential, proximity to employees, fibre optic cable and cheap land or rent (Nixey, 2000, p. 15-17). Surprisingly, amenities such as high design standards, outdoor recreation, transit, commercial services and restaurants all out-ranked proximity to other high-tech firms, the supply chain or customers (Table 1).

Table 1. Comparison of High-tech Firm Locational Preferences and Firm Locations

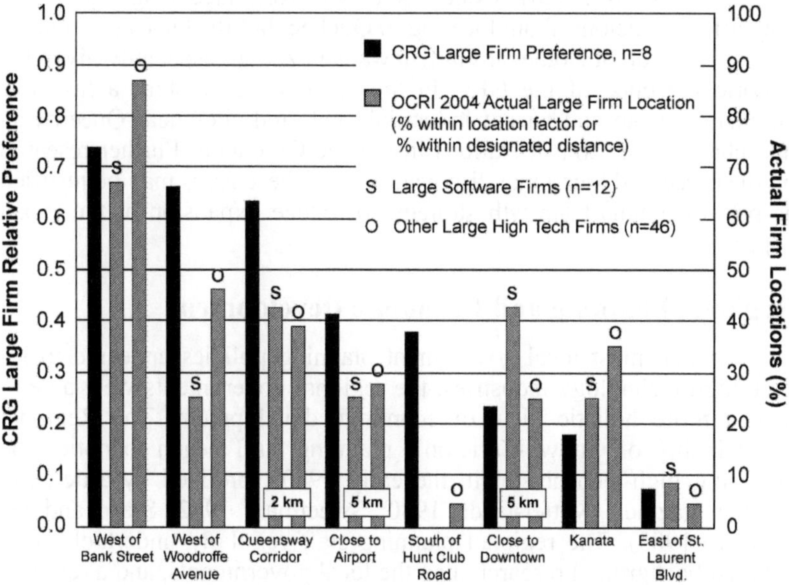

Source: Nixey, 2000; Ottawa, 2004; Drawn by John Kozuskanich.

Small firms had slightly different concerns, with cheap land/rent, expansion potential, parking, an understanding landlord, transit and design quality topping their list. Proximity to customers, suppliers and other tech firms ranked below these factors, but were ranked somewhat more important than the large employers. However, it appears that the competitive factors featured in Michael Porter's (1998) diamond of context, input conditions, demand conditions and related industries, don't appear to be the major location drivers at the micro level within the Ottawa region.

The high-tech executives had some clear location preferences, which may be partially driven by historical patterns. The most important geographic area for companies was "close to home base," perhaps explaining some of the patterns of start-up companies. Both small and large companies had a strong bias towards locations in the southwest and the Queensway corridor, but small firms had a higher interest in location close to their downtown customers, especially in software consulting (Nixey, 2000).

Less than 10% of the executives would consider locations in the east end of Ottawa or Québec, which were ranked the lowest in both surveys (Nixey, 2000, p. 12-20). Concerns about workplace language issues may affect decisions about locating in Québec, but the bias against lands east of St. Laurent Boulevard in Ottawa is puzzling, especially given the historic presence of the NRC in this area. Other eastern areas have abundant, cheap and vacant industrial land, and sites near Queensway interchanges and transit stations inside the Greenbelt. Further research on the perceived locational disadvantages of these areas may be justified as part of a Smart Growth strategy to reduce expansion at the urban edges.

Regional Planning and Economic Development

Although most local government planning policies appeared to ignore high-technology industries, the regional governments were able to take a more holistic view of economic development. The Regional Municipality of Ottawa-Carleton's planning staff began to notice the emerging high-tech cluster in the early 1980s, prodded by articles in local magazines (Mittelstaedt, 1980; Sweetman, 1982; Steed and De Genova, 1983). The regional planning staff regularly undertook more detailed background research than the local governments, and a regional employment survey (Ottawa-Carleton, 1987) first mentioned the growing concentration of high-technology industries.

The regional governments were looking for means to diversify the economy of the National Capital Region in the late 1980s and mid-1990s because the federal government was actually contracting as a direct employer during this period. The Conservative government elected in 1984 moved vowed to give civil servants "pink slips and running shoes." They were followed by Liberal governments elected in 1993 and 1997, which reduced employment while eliminating the federal budget deficit. Federal employment in the Ottawa-Hull Census Metropolitan Area declined from 115,590 to 93,640 over 1986-1996 (Statistics Canada, 1986; 1996), the first decrease in the 20th century (Taylor, 1986, Appendix IX).

The decline in the importance of federal employment occurred while the RMOC became more activist on planning issues. Regional councillors were also local councillors from 1968 to 1994 and their dual allegiance may have compromised their effectiveness on some regional planning issues. The Conservative provincial government elected in 1995 downloaded responsibility for transit funding and social services to the regional government. The late 1990s therefore saw a directly elected regional council responsible for both the costs and benefits of urban development for the first time.

Regional council saw economic diversification as a means to 'grow the pie' and maintain the region's quality of life. But the regional government did not attempt to seize control of the three growth sectors they identified: tourism, high technology and bioscience. Instead, they supported establishment of partnerships and networks led by the private sector through the Ottawa Tourism and Convention Authority (OCTA) and the Ottawa Centre for Research and Innovation or OCRI (Ottawa-Carleton, 1992, p. 1-4). The RMOC supported OCRI with facilities, research and financial assistance, but strategic direction came from the stakeholders in the organisation. OCRI's board included the chief executives of all the research labs, and senior executives of the largest technology firms. It fostered networking, human resource-fostered networking, human resource development and other factors for 'collaborative advantage' (Gertler and Wolfe, 2004; Chamberlain and de la Mothe, 2003; de la Mothe and Mallory, 2003; Leibovitz, 2000).

The background studies for the 1997 regional official plan included an economic development strategy based on these principles (Ottawa-Carleton, 1999). It contained only two paragraphs and one recommendation regarding the region's largest employer, the federal government. The remaining 23 recommendations of the 28-page report were to implement initiatives in high technology, life sciences and tourism. Perhaps the most important recommendation was that Regional Council should help establish The Ottawa Partnership (TOP), including the chairs of OTCA, OCRI, the Ottawa Life Sciences Council, the Ottawa Economic Development Corporation, the Board of Trade, and the universities.

The 1997 Regional Official Plan called for curbing urban sprawl by redevelopment inside the Greenbelt (Ottawa-Carleton, 1997). But the low-density, car-oriented preferences of high-technology industries may undercut this commitment to compact development. For example, the owners of the property surrounding the Corel Centre hockey arena on the western edge of Kanata applied for the urban growth boundary to be extended to permit high-technology business park development on their lands in the late 1990s. Despite evidence that the region's overall supply of industrial land was strong and that there was little complementarity between an entertainment venue and high-technology locations (Nixey, 2000), Regional Council approved the extension, based partly on optimistic forecasts of further geometric growth in high-technology employment.

Regional and local governance and planning were combined by the 2001 provincial reorganisation, which amalgamated 11 local governments and the Region of Ottawa-Carleton into a new City of Ottawa. The Ottawa Partnership was influential in the agenda for the new City, and sponsored important background studies for the Ottawa 2020 plan (ICF Consulting, 2000; 2002; Doyletech *et al.*, 2002). These studies were the foundation of *Ottawa 20/20 – Economic Strategy*, one of the five major components of the City's growth management strategy. The *Economic Strategy* had twelve key recommendations for action:

1. Improve two-way communication with the business community.
2. Implement a targeted business visitation programme.
3. Address workforce shortages, underemployment and training issues.
4. Extend broadband service to every Ottawa address.
5. Increase commercialisation of research and technology.
6. Leverage Ottawa's business incubators – not just for technology but for all sectors.
7. Help Ottawa businesses market their products and services better.
8. Create an infrastructure to manage and support industry cluster collaboration.
9. Promote increased commerce between the industry clusters and local businesses, and between urban and rural businesses.
10. Promote and strengthen Ottawa's main street business areas.
11. Market and promote Ottawa domestically and internationally as an ideal location for people, tourism, companies and investment.
12. Move forward the expansion of the Ottawa Congress Centre (Ottawa, 2003a).

The City's new comprehensive land use plan (Ottawa, 2003b) complements the economic strategy. The Quebéc government imposed a similar regional reorganisation in 2002. The Outaouais local governments were consolidated into two municipalities – urban and rural – that are also preparing new plans promoting employment diversification.

Predictions that the technology sector would surpass the federal government as the largest employer in the National Capital Region by the end of the century proved optimistic. The technology sector, especially the telecommunications cluster, contracted sharply in 2001-2002, when over 20,000 jobs were lost. But it appears that many of these highly skilled employees stayed in the region, some started new firms, and employment has picked up again to perhaps 67,000 jobs in 2003 (Ottawa, 2003a, p. 27). The 'dot.com meltdown' was not as severe as was expected for Ottawa because its technology sector embraces microelectronics and software in addition to the troubled telecommunications cluster. There were surprisingly few bankruptcies and the remaining companies made tactical choices to survive the downturn (Madill, Haines and Riding, 2004).

Ottawa's economy did not collapse during the high-tech contraction of 2001. Housing prices continued to climb, new housing construction did not decline and employment actually increased (Madill, Haines and Riding, 2004). An expansion of federal spending in the late Chrétien era picked up the slack, just as the high-tech sector did in the late 1980s/early 1990s. The National Capital Region was fortunate to have two employment sectors, government and high technology, that ran in opposite cycles.

Conclusions: From Red Tape to Blue Chips?

The federal government's laboratories and early assistance to Northern Electric certainly helped establish the first round of high-technology employment in Ottawa (Harrison, Cooper and Mason, 2004; Smart, 2004; Ghent-Mallet, 2002). But, Ottawa's balanced economy and surprisingly strong high-technology sector are not the direct result of deliberate regional planning by government. However, they are perhaps unanticipated benefits of the century-long effort to transform a lumber town into an attractive national capital. Planning a national capital in the late 20th century required some attention to creating a city that was beautiful, functional and included social, cultural and recreational facilities that were attractive to highly educated civil servants and diplomats (Gordon, 2006). This type of planning often has benefits in increasing cultural tourism activity in a capital city.

The unique investments made by the federal government to improve the national capital also improved the quality of life sufficiently that many highly educated research and development staff could be attracted to the federal labs or Nortel. More importantly, they remained in the region after they left their original employer, sometimes starting their own firms. The federal government's presence has helped to create Canada's most highly educated metropolitan labour force, which is believed to be a key component for developing a high-tech cluster (Gertler, Florida *et al.*, 2002). Washington, DC, has a similar profile in the United States and has also been an attractive location for communications and high-technology employers like MCI (Florida, 2002; Feldman, 2001).

In Ottawa, as in Washington, the quality of life improvements planned and co-ordinated by the national government to make an attractive capital city may have had the unanticipated benefit of making a place that was attractive to high-technology workers. The Ottawa-Gatineau region has Canada's longest history of urban planning because it is the national capital. The nature of the planning has changed over the past century as the regional and local governments have become more engaged in the process (Table 2).

Table 2. Comparison of High-tech Firm Locational Preferences and Firm Locations

	1900 – 1930	1945 – 1960	1970 – 1990	1990 +
Planning Objectives	Federal: beautification Local: basic services	Federal: capital building Local: suburban expansion	Local/Regional: suburban expansion	Federal: national unity Regional: economic development
Planning Techniques	Comprehensive plans	Regional planning	Local land use plans Regional land use plans	Strategic planning Collaboration - building Smart Growth
Key Actors	OIC FDC	FDC NCC City of Ottawa	RMOC/CUO Local governments	RMOC/CUO 'New' City of Ottawa OCRI/OCTA
Key Infrastructure	Streetcars Railways Water supply Telephone	Railways Expressways Sewers	Expressways Busways Airports	Busways Light rail Fibre optic cable

The last decade has seen closer attention to economic development initiatives, but this approach was incorporated into planning policy documents just in time for a serious decline in high-technology employment. But, the city/regional government's efforts to build public-private partnerships have thickened collaborative institutions and may have helped maintain a foundation for future growth. The planning policies that William Lyon Mackenzie King initiated to serve diplomats and tourists may have made Canada's capital an attractive location for the next generation of highly educated knowledge workers.

References

Best, M. (2001). *The new competitive advantage*. Oxford: Oxford University Press.

Chamberlain, T., de la Mothe, J. (2003). Northern light: Ottawa's technology cluster. In *Clusters old and new*, D.A. Wolfe (ed.). Kingston: McGill-Queen's University Press, 8, 213-234.

Community Planning Association of Canada. (1952). *The city nuclear. A regional pattern for urban growth*. A presentation made by the Community

Planning Association of Canada to the Fifteenth Annual Conference of the Canadian Federation of Mayors and Municipalities.

Community Planning Association of Canada. (1966). *What's new in planning: report for 1965*. No.8. A bulletin for citizens in the National Capital Region. Issued by the National Capital Region Branch. Ottawa: Community Planning Association of Canada.

Community Planning Association of Canada. (1972). A further report on long-range planning for CPAC. Prepared for the National Council.

de la Mothe, J., Mallory, G. (2003). *Industry-government relations in a knowledge-based economy: The role of constructed advantage*. Ottawa: University of Ottawa.

Dissart, J.D., Deller, S.C. (2000). Quality of life in the planning literature. *Journal of Planning Literature, 15* (1), 135-161.

Donald, B. (2001). Economic competitiveness and quality of life in city regions: compatible concepts? *Canadian Journal of Urban Research, 10* (2), 259-274.

Donald, B., Morrow, D. (2003). *Competing for talent: implications for social and cultural policy in Canadian city-regions*. Ottawa: Heritage Canada.

Dosi, G. (1998). Sources, procedures and microeconomic effects of innovation. *Journal of Economic Literature, 26*, 1126-1171.

Doyle, D. J. (1991). *From white pine to red tape to blue chips; how technology based industries can provide continued prosperity for the Ottawa-Carleton Region*. Ottawa: Ottawa-Carleton Economic Development Task Force.

Doyletech Corporation. (2002). *The family tree of Ottawa-Gatineau high technology companies 2002*. Poster.

Doyletech Corporation. (2003). *Humble beginnings: a pictorial history of the Ottawa-Gatineau high technology industry*. Poster.

Doyletech Corporation, McSweeney Associates Management Consultants, Market Research Corporation. (2002). *Serving Ottawa: economic strategies to develop Ottawa's local economy*. Ottawa: City of Ottawa.

Eggleston, W. (1978). *National research in Canada: The NRC 1916-1966*. Toronto: Clarke, Irwin & Company Limited.

Elliot, B.S. (1991). *The city beyond: a history of Nepean, birthplace of Canada's capital 1792-1990*. Nepean: City of Nepean.

Federal District Commission. (1958) *Annual report 1958*. Ottawa: FDC.

Federal Plan Commission of Ottawa and Hull (FPC). (1916). *Report of the federal plan commission on a greater general plan for the cities of Ottawa and Hull: 1915*. Ottawa: FPC.

Feldman, M.P. (2001). The entrepreneurial event revisited: firm formation in a regional context. *Industrial and Corporate Change, 10*, 861.

Feser, E.J. (2003). What regions do rather than make: a proposed set of knowledge-based occupation clusters. *Urban Studies, 40* (1), 1937-1958.

Florida, R., Gates, G. (2001). *Technology and tolerance: the importance of diversity to high-technology growth*. Washington, DC: The Brookings Institution.

Florida, R. (2002). *The rise of the creative class*. New York: Basic Books.

Florida, R. (2005). *The Flight of the Creative Class: The new global competition for talent*. New York: Harper Business.

Gertler, M.S. (2001). Urban economy and society in Canada: flows of people, capital and ideas. *Isuma*, *2* (3), 119-130.

Gertler, M.S., Florida, R., Gates, G. and Vinodrai, T. (2002). Competing on creativity: placing Ontario's cities in North American context. A report for the Ontario ministry of enterprise, opportunity and innovation and the institute for competitiveness and prosperity.

Gertler, M.S. and Wolfe, D.A. (2004). Local social knowledge management: community actors, institutions and multilevel governance in regional foresight exercises. *Futures*, *36*, 45-65.

Ghent-Mallet, J. (2002, October). *Silicon valley north: the formation of the Ottawa innovation cluster*. Ottawa: Information Technology Association of Canada.

Glaeser, E.L. (2000). The new economics of urban and regional growth. In G. Clark, M. Feldman and M. Gertler (eds.). *The Oxford handbook of economic geography*. Oxford: Oxford University Press, 83-98.

Glaeser, E.L. (2004). Review of Richard Florida's The Rise of the Creative Class. Retrieved on July 14, 2006 from http://post.economics.harvard.edu/faculty/glaeser/papers/Review_Florida.pdf.

Gordon, D.L.A. (1998). A city beautiful plan for Canada's capital: Edward Bennett and the 1915 plan for Ottawa and Hull. *Planning Perspectives*, *13*, 275-300.

Gordon, D.L.A. (2001a). William Lyon Mackenzie King, town planning advocate. *Planning Perspectives*, *17* (2) 97-122.

Gordon, D.L.A. (2001b). Weaving a modern plan for Canada's capital: Jacques Gréber and the 1950 plan for the national capital region. *Urban History Review*, *29* (2), 43-61.

Gordon, D.L.A. (2002a). Frederick G. Todd and the origins of the park system in Canada's capital. *Journal of Planning History*, *1* (1), 29-57.

Gordon, D.L.A. (2002b). Ottawa-Hull and Canberra: implementation of capital city plans. *Canadian Journal of Urban Research*, *13* (2), 1-16.

Gordon D.L.A., Osborne, B.S. (2004). Constructing national identity: confederation square and the national war memorial in Canada's capital, 1900-2000. *Journal of Historical Geography*, *30* (4), 618-642.

Gordon, D.L.A. (ed.). (2006). *Planning twentieth century capital cities*. London: Routledge.

Granovetter, M. (1985). Economic action and social structure: the problem of embeddedness. *American Journal of Sociology*, *91*, 491-510.

Gréber, J. (1950). *Plan for the national capital, general report*. Ottawa: National Capital Planning Service.

Harris, R. (2004). *Creeping conformity*. Toronto: University of Toronto Press.

Harrison, R.T., Cooper, S.Y., Mason, C.M. (2004). Entrepreneurship activity and the dynamics of technology-based cluster development. *Urban Studies, 41* (5/6), 1045-1070.

ICF Consulting. (2002). *Innovation Ottawa: a strategy for sustaining economic generators*. Ottawa: The Ottawa Partnership.

ICF Consulting. (2000). *Choosing a future: a new economic vision for Ottawa.* Ottawa: The Ottawa Partnership.

Jacobs, D. (2004). City of Nepean, former planner, City of Ottawa Director, Planning, Environment & Infrastructure Policy Branch, interview with D. Gordon and J. Kozuskanich, 15 June 2004.

Kanata, City of. (1999). *Kanata – an economic profile. Spreading the good news to the rest of the world.* Kanata: Economic Development Office.

King, W.L.M. (1900). William Lyon Mackenzie King diary. Ottawa: Library and Archives Canada.

Leibovitz, J. (2000). *Barriers to collaborative city-region governance: institutions and economic governance in Silicon Valley North.* Unpublished Doctoral dissertation, University of Glasgow, UK, Department of Urban Studies.

Lundvall, B-A (ed.). (1992). *National innovation systems: towards a theory of innovation and interactive learning*. London: Pinter.

MacKay, R. (2004). Former City of Kanata planner, City of Ottawa manager of business initiatives, interview with D. Gordon and J. Kozuskanich, 16 October 2004.

Madill, J.J., Haines, G.H., Riding, A.L. (2004). *Advanced technology firms: strategies for surviving tough times*. Ottawa: Carleton University.

Markusen, A. (2004). Targeting occupations in regional and community economic development. *Journal of the American Planning Association, 70* (2), 53-69.

Mittelstaedt, M. (1980). Ottawa: the new high-tech haven. *Canadian Business, 53*, 43-46, 82-87, 105.

National Capital Commission. (1970). *Annual report 1970*. Ottawa: NCC.

Nixey, D. (2000). *Corel centre lands: justification of need.* Ottawa: Region of Ottawa-Carleton, Department of Planning and Development Approvals.

O'Neill, J. (2005, October 10). A love affair with Ottawa. *Ottawa Citizen,* p. A1.

Ottawa, City of. (1953). *Official plan of the Ottawa planning area.* Ottawa: Ottawa Planning Area Board.

Ottawa, City of. (2000). *History of population projections in Ottawa 1915-2001*. Ottawa: Development Services Report.

Ottawa, City of. (2003a). *Ottawa 20/20 economic strategy*. Ottawa: Planning and Building.

Ottawa, City of. (2003b). *Ottawa 20/20 official plan.* Ottawa: Planning and Building.

Ottawa-Carleton Regional Municipality. (1974). *Official plan, Ottawa-Carleton region. Final draft recommended by regional planning committee to executive committee & council*. Ottawa: Planning Department.

Outaouais, Communauté Régionale de l'. (1976). *Schéma d'aménagement du territoire*. *CRO:* Regional Planning Department.

Ottawa-Carleton, Regional Municipality. (1987). *Employment in Ottawa-Carleton 1976-1986*. Ottawa: Planning Department.

Ottawa-Carleton, Regional Municipality. (1992). *Partners for the future: a strategic economic vision for Ottawa-Carleton*. Ottawa: Economic Development Task Force.

Ottawa-Carleton, Regional Municipality. (1997). *Official plan – region of Ottawa-Carleton*. Ottawa: Planning Department.

Peck, J. (2005). Banal urbanism: creativity as scalar narrative. Paper presented at the Studies in Political Economy Conference, 'Towards a political economy of scale.' Toronto: York University.

Ploegaerts, L. (1973). *Étude critique et comparative des plans d'urbanisme et de développement municipaux préparés depuis 1967 sur le Territoire de la Communauté régionale de l'Outaouais*. Ottawa: University of Ottawa Geography and Regional Planning.

Ploegaerts, L. (1976). *Étude comparative des règlements de zonage sur le territoire de la Communauté régionale de l'Outaouais*. Ottawa: University of Ottawa Geography and Regional Planning.

Porter, M.E. (1998). Clusters and the new economics of competition. *Harvard Business Review, November-December*, 77-90.

Smart, J.D. (2004). *A City and a technology: the role of government in the growth of the technology industry in Ottawa, Canada to 1984*. Unpublished PhD Dissertation. Kingston: Queen's University, Department of History.

Statistics Canada. (1986). Census, Ottawa-Hull census metropolitan area.

Statistics Canada. (1996). Census, Ottawa-Hull census metropolitan area.

Steed, G.P.F., DeGenova, D. (1983). Ottawa's technology-oriented complex. *Canadian Geographer*, *27* (3), 263-278.

Sweetman, K. (1982). Ottawa also is our high-tech capital. *Canadian Geographic*, *102*, 20-31.

Taylor, J.H. (1986). *Ottawa: an illustrated history*. Toronto: J. Lorimer.

Taylor, J.H. (1989). City form and capital culture: remaking Ottawa. *Planning Perspectives*, *4*, 79-105.

Thomas, D. (1983). *Knights of the new technology: the inside story of Canada's computer elite*. Toronto: Key Porter Books.

Timusk, C. (1976). Kanata: a new community approaches its tenth year. *Journal of Urban and Environmental Affairs*, *8* (3), 222-232.

Todd, F.G. (1903). *Preliminary report to the Ottawa improvement commission*. Ottawa: OIC.

Part III

'Growing' the Cluster
Idea Farming and Innovation Strategies for Economic Development

CHAPTER 5

Steering on Black Ice

The Continuing Search for Sustainable Livelihoods in the Ottawa Tech Sector

Edward T. JACKSON and Rahil KHAN

School of Public Policy and Administration, Carleton University

When you hit a patch of invisible 'black ice' on the road, it is usually too late to do anything other than steering away from the ditches and oncoming cars – and hoping for the best. If you slide off the road, you will have to rely on the goodwill of other drivers to help you until the tow-truck arrives and you can resume your journey. In the final analysis, the only real guarantee you can succeed in steering on black ice is for the odds to be in your favour that the community will come to your aid.

Introduction[1]

In most countries of the developed world, the jury is still out on whether technology clusters are capable of generating sustainable livelihoods for large numbers of households. In technology centres across North America and Europe, the question remains open – and serious. How does a technology cluster rebuild itself after a radical collapse? How do technology workers, their households and the communities in which they live develop the *resilience* necessary to prevail in the face of such adversity? What strategies and tools can be created and effectively applied in support of this struggle for sustainable livelihoods? This chapter examines Ottawa's recent experience in coping

1 The authors are grateful to the following individuals for their advice and assistance in the preparation of this paper: Bob Barclay, Lou Barsoney, Gary Davis, Josh Korn, Jim McQuaid, Alex Newcombe, Paul Putinski, Rick Radko, David Sproule and Sonja Vanek. Marianne Wilkinson, David Large and Sandra Huntley provided helpful comments on an earlier draft. The 'black ice' metaphor originated in a news story by Juliet O'Neill.

with tech-cluster adversity – in trying to steer on the 'black ice' of uncertainty and volatility in the tech sector. The lessons of this experience may be of use to technology clusters elsewhere.

Picking Up the Pieces After the Tech Wreck

Before the technology bubble burst in 2001, and powered by computers and telecommunications, the sector delivered hundreds of thousands of well-paying jobs, notably to highly skilled engineers and programmers, across North America and Europe. Along with these knowledge workers also came hundreds of thousands of lower skilled employees, including secretaries, clerks and laboratory assistants. In cluster centres like Palo Alto, Boston and Austin, local economies boomed through big-ticket purchases on homes and cars by tech employees. Economic multipliers pulsed through these centres, enabling the services, retail, financial, construction and real estate sectors to flourish. Tax revenues filled the coffers of municipalities. Charitable donations increased sharply.

Then, the tech wreck hit. The sector unravelled, and shrunk with astonishing speed. Large, anchor companies in telecom, photonics, software and semi-conductors shed tens of thousands of workers, high – and low-skilled alike. Smaller, supplier firms lost their markets and failed. Venture capital investment in the sector fell dramatically. The volatility and unpredictability of the tech sub-sectors – together with the non-transferability of some technology skills *across* those sub-sectors – made it extremely difficult for knowledge workers to anticipate and move toward job opportunities. In contrast to the buoyant optimism several months earlier, tech workers suddenly faced the spectre of long-term unemployment, and many experienced depression. Financial and emotional stress ate away at family stability. Homes were mortgaged or sold, and households tapped their core savings to cope with the downturn. Some disillusioned knowledge workers abandoned the sector altogether, migrating to lower-paid professional or service jobs. Without tech-driven multipliers, other sectors faltered. Tax revenues and charitable donations shrank markedly, putting further pressure on governments and charities to respond to the social consequences of the 'wreckage.'

This pattern describes the Ottawa experience. From a high of more than 70,000 workers during the boom, Ottawa's technology cluster now employs about 45,000 people according to Statistics Canada. But not only was there a dramatic drop in jobs in absolute terms, there was also a decline in the *quality* of employment. Much more tech employment in

Ottawa now is part-time, lower paid, with fewer benefits, and often in risky start-ups.[2] In 2004, only five of 33 publicly traded tech companies in Ottawa posted share prices in the double digits. A dozen high-profile companies shut their doors. Venture capital investments were minimal and slow. Semi-conductor sales were flat.

With a high Canadian dollar, those technology firms still active in the region sought to limit their operating costs by not taking on new staff. The 10% office vacancy rate was the highest in seven years, with the most elevated levels located in the west-end suburb of Kanata. Tight municipal finances resulted in sharp budget cuts to community agencies, and many charities experienced loss of income from a drop in donations. Additional pressure was placed on the local community foundation and other donors to compensate for this funding loss, which they could only fulfill to a limited extent. Residents were spending more of their income than two years earlier on essentials like housing, furnishing, savings and taxes. Perhaps most disturbingly, there was unprecedented demand in 2004 on food-bank services in the region. Fully one-fifth of Ottawa-area food-bank users hold a post-secondary degree, and many of these are likely former knowledge workers (Kristal-Schroder, 2004). The local media extensively chronicled the region's economic downturn and its effects (see, especially, Bagnell, 2005a, b; Goff, 2004a, b, c; 2005 a, b, c, d, e; Hill, 2004a, b, c, d; Hum, 2005; Mayeda, 2005, 2004a, b, c; O'Neill, 2004, 2005; Sutcliffe, 2004, 2005).

One of the least impressive features of the Ottawa experience was the way in which politicians, business leaders and some media editors tended to minimise or obscure the scale and negative effects of job loss in the tech sector. Apparently not able to break their habit of boosterism honed during the tech boom, these public voices would accept a number no higher than 5,000 net unemployed tech workers. Aided by the efforts of the Ottawa Centre for Research and Innovation, the City's economic arm, to present the most positive tech employment numbers possible, and in the absence of comprehensive, ground-level survey data, this figure was maintained in public discussions for several years.[3] However,

2 Recent national research by Statistics Canada suggests that the most devastating effects of the downturn were concentrated in information and communications technology (ICT) manufacturing. In contrast, the numbers of businesses in ICT services remained relatively stable. And the sector continues to attract a relatively large number of new businesses each year, in spite of the downturn (see Beckstead and Brown, 2005). It is likely that, in Ottawa at least, these new entrants are primarily start-ups that have emerged following the shedding of jobs by the largest companies.

3 On the size of the current labour force in the technology sector, OCRI puts that figure as of early 2005 at 67,800 – although this includes non-technology workers and support staff in tech firms. In contrast, Statistics Canada, using different criteria, esti-

as activists among unemployed tech workers organised themselves, and began to speak out publicly, a more realistic and accurate discourse eventually emerged. In early 2005, a full five years after the peak of the bubble, an Ottawa Citizen article used Statistics Canada data to confirm that the tech sector had lost 27,000 net jobs since the downturn began. Ottawa's largest anchor firm, Nortel, alone shed 15,000 jobs during the period. This perspective lent new credibility to activist claims that there are 10,000 to 15,000 unemployed or underemployed technology workers in Ottawa today.

While this picture obviously is not pretty, Ottawa has some key assets that are helping the region navigate through this challenging period. Being a capital city, Ottawa is home to the headquarters of most federal-government departments and agencies. When the bubble burst, some tech workers quickly moved over to work in the public sector, with its job security and attractive salaries and benefits. Moreover, government is an important purchaser of the products and consulting services of computer, software and telecom companies. In particular, terrorism and the Iraq war have fuelled government procurement in tech-related security and defence products. Further, a robust construction programme by the federal government has strengthened that sector. At the same time, as the tragic events of 9/11 recede in time, Ottawa's tourism sector, which relies on a mobile American clientele, is beginning to revive. Finally, the region is home to some 1,600 vibrant small and medium-sized businesses, half of which hired new staff in 2004.

Still, the Ottawa economy remains stalled in low gear, and tech prospects are not particularly bright. Layoffs continued in Ottawa's larger firms in 2005, notably Mitel, Nortel, and Alcatel. Forecasts for the middle of the decade call for modest growth in retail but minimal growth in manufacturing, which in Ottawa is primarily technology-based. Some analysts project that the Ottawa tech sector can not fully rebound until 2010 – a very long time for unemployed and underemployed knowledge workers and their families. As the region continues to pick up the pieces after the tech wreck, all sectors – business, government, labour and the non-profit sector – must recommit to work even harder to find a viable way forward. It turns out that rebuilding the

mates that technology employment in the region reached only 45,500 by that date (Goff, 2005e). Both groups agree, though, that there has been a dramatic fall in tech jobs since the bubble burst. Yet, presumably to promote the region, OCRI's president publicly predicted that tech-sector employment will climb to 100,000 jobs by 2008. This prediction was greeted with wry scepticism in many quarters (see Sutcliffe, 2005). It also served to maintain the ambiguity in public discourse on this issue.

technology cluster in Ottawa is not actually a sprint. It is really more of a marathon.

So, who has been doing what? In the non-profit sector in Ottawa, there is widespread agreement on the importance of taking an asset-based rather than a needs-based approach to community revitalisation. That is, the economic, social, environmental and human assets of all segments of the community must be mapped and mobilised in order to build an effective local development and poverty-reduction strategy. And most non-profits would consider knowledge workers to be assets. However, while they have tried in some ways to respond to the needs of unemployed technology workers, nonprofits like the Community Foundation, the Social Planning Council, the Ottawa Community Loan Fund and the Ottawa Food Bank have been necessarily preoccupied with the City's budget cuts to community groups. Of necessity, these organisations have concentrated their efforts on the very needy and hard-core poor, among whom are new immigrants, single-parent women, disaffected youth and the homeless. At the same time, links between the non-profit sector and the technology labour force were never strong during the boom period, and these parties have not been able to easily forge a close alliance in the new circumstances (see Jackson and Khan, 2003). In the aftermath of the downturn, however, local community resource centres, particularly these in the western part of the city, began establishing working groups and programmes targeted specifically to the needs of technology workers.

In contrast, for most of the period since the bubble burst, the Ottawa Centre for Research and Innovation (OCRI), the municipality's economic development arm, has tried to promote the tech sector as if it were business as usual. In fact, the tendency of the OCRI leadership has been to continue to sell the region internationally as if the boom was still on, a position that frequently rang hollow both at home and abroad. Trade missions to American, Asian and European cities to attract companies to set up operations in Ottawa, and to promote joint ventures with local firms, remained an important part of OCRI's work. However, important as they were, any modest successes in these areas were not, to say the least, sufficient to replace the critical mass of businesses and jobs that were eliminated in the downturn. At the same time, OCRI encouraged its Entrepreneurship Centre to assist the many new technology start-ups that emerged out of the tech wreck. This was useful but, again, came nowhere near to replacing the thousands of lost jobs, either in quantity or quality.

A New Adjustment Tool The Ottawa Talent Initiative

However, as a result of the organising efforts of former tech workers themselves, OCRI eventually agreed to host a new project – the Ottawa Talent Initiative (OTI) – to promote more effective and broad-based training and adjustment of this labour force. OTI began operations in 2004. As its mission statement says, OTI is "a grass roots organization composed of unemployed technology workers and community volunteers working together to create positive change for the unemployed and underemployed technology workers in the Ottawa area" (Ottawa Talent Initiative, 2006). A review of this initiative provides insights and lessons on the challenges faced by a technology cluster seeking to rebuild itself, as well as the possibilities of specific rebuilding strategies. We turn to this case study now.

Origins

When large-scale layoffs began in early 2001, laid-off workers formed a variety of peer-networking groups to support each other in job searches, business start-ups and mentoring. By 2003, however, group members realised they faced a much larger problem and needed to coordinate their efforts. In mid-2003, a town hall meeting on the tech downturn was organised by the Western Ottawa Community Resource Centre (WOCRC). In October 2003, a large networking group, Cabin Global, hosted another public event intended to generate hope among laid off workers. Expecting 50 participants, the organisers were surprised when 400 people showed up. Newspaper articles were now regularly bringing the negative effects of the tech layoffs to public attention (O'Neill, 2003a, b). Although politicians were still not drawn to the problem, some did pay attention when approached by key stakeholder groups.

Subsequent discussions among key peer groups representing unemployed tech workers resulted in the establishment, in January 2004, of the Ottawa Talent Initiative. The OTI name was chosen for those groups that had been meeting since November 2003. A team was established, varying from eight to 12 members at any one time, to prepare a Community Action Plan (CAP) for an adjustment strategy for the region. The next month OTI held a large public meeting, the Ottawa Talent Forum, that drew some 500 unemployed and under-employed technology workers, as well as leaders from community agencies, government and business. Input into the Action Plan was sought from the Forum attendees. The Ontario Ministry of Training, Colleges and Universities provided 15,000 Cdn$ to OTI to develop the Plan, routing these funds through OCRI. A team of two facilitators was hired to assist the team in preparing the Community Action Plan. The Action Plan initially suggested 18 key strategies as a starting point, plus additional strategies that could be developed over time.

The three levels of government were not directly involved in OTI's establishment, although they did play a role in the organisation's development. For instance, the City of Ottawa contributed a staff member to the OTI Steering Committee and OCRI also committed an individual as a key Steering Committee member. The Province also deployed a staff member who later joined the Steering Committee and helped in providing the 15,000 Cdn$ in funding to develop the Action Plan. One meeting was held with provincial and federal staff present to discuss funding. However, that was the only tangible federal involvement in the start-up process.

On the basis of the CAP, the Ministry then provided OTI with an operating grant of 180,000 Cdn$ under its Adjustment Advisory Programme (AAP), which helps local stakeholders deal with the immediate effects of plant closures and down-sizings, usually in traditional, 'rust-belt' industries. OTI's top two priorities were, first, to establish an Action Centre as a focal point from which to address the needs of highly skilled unemployed and underemployed technology workers and, second, to create employment for them. AAP funds can be used for honoraria for adjustment committee members – OTI qualifies as an adjustment committee – as well as for outside consultants and other programme costs. The OTI Steering Committee was composed of unemployed volunteers from its 20 affiliated community-based peer-network groups, plus supporters of the Initiative from governments and other community organisations. However, the Committee members did not receive remuneration for their time and expenses. The OTI Action Centre opened its doors in July 2004. Within the Action Centre, there is a limited number of paid and mostly part-time individuals who operate the Centre.

In addition to organising a second Talent Forum in early 2005, OTI formed a number of Task Forces to implement the CAP priority activities. It is also currently preparing a two-year, 2-million Cdn$ programme proposal that would pursue more 'demand-side' initiatives, particularly through business contacts and by working with local universities and colleges on commercialisation activities. For instance, one idea is the establishment of business oriented R&D centres that would help generate jobs and businesses by accelerating the commercialisation of new tech innovations by laid-off workers. Advised by politically-savvy leaders, the Initiative is focusing its lobbying for these funds on the federal and provincial Liberal-Party caucuses given that Liberal governments are in power in both jurisdictions.

The Community Action Plan

Released in May 2004, the CAP is meant to serve as a framework for job-seekers, economic-development agencies and all levels of government to coordinate action to address the problem of persistent unemployment and underemployment among technology workers. Building on previous municipal policy and planning documents, the Plan focuses on three strategies: helping to grow jobs in the Ottawa region, gaining immediate support for unemployed tech workers, and providing assistance to unemployed and underemployed workers to help them make the transition to other sectors. The CAP identifies a list of proposed projects ranging from extending employment insurance benefits to developing a high-tech specific view of economic forecast data. Overall, while the Plan served the purpose of securing the provincial operating grant, OTI is evolving beyond the CAP framework.

Structure and Operations

The OTI Steering Committee no longer exists, having fulfilled its early-stage functions. At the heart of the Initiative are unemployed technology workers, engaged in the activities emerging out of the OTI Action Centre. While OCRI continues to serve as the corporate vehicle for OTI (providing, among other things, administrative and financial services), the Initiative now operates independently and is considering incorporating as a non-profit entity. OTI's operating principles include stimulating economic growth, reducing the social and economic effects on unemployed individuals, identifying ways to sustain the tech talent pool in the future, and continuously evaluating and improving its progress. To achieve these principles, OTI's main strategy is to get people involved – both those directly affected by the downturn and those in the broader community. It is, in essence, a vehicle for worker and citizen mobilisation.

The Initiative's activities are primarily undertaken through task forces that are organised under key themes in the CAP. For example, under the theme of 'Creating Business Opportunities,' it was envisioned that certain task forces would work on building a database of individuals' skills and coordinates, forming teams to bid on government requests for proposals, conducting research on labour market and workforce growth trends, and establishing links with CEOs and human resource directors in local companies as well as with recruiters and head-hunters.

Under the theme of 'Individual Development,' task forces would focus on gaining security clearance for individuals, accessing entrepreneurship assistance, building relationships with the academic community, and setting up a speakers' series on a variety of topics.

These task forces require significant effort to set up and manage. Consequently, only a small number of the planned task forces are actually up and running. The CAP did not specify that each initiative requires a Task Force, but rather identified this as one mechanism by which to implement the plan. Presently, several strategic areas are grouped under one Task Force, while other initiatives are run directly by the Centre rather than a Task Force. For instance, the Centre provides professional examination preparation in project management; a speakers series; and mini courses (e.g., technical writing, coaching, mentoring and individual help). A Monday morning coffee drop in, usually with a special speaker, helps to draw people out of their homes and prepare them psychologically to begin their week of job searching and other employment-related activities.

Outstanding Gaps and Issues

The most pressing gap for OTI is continued funding. There is a growing constituency that wants the organisation to survive and even expand. The Initiative's activities are already making a difference in people's lives and succeeding in obtaining jobs for those involved with the Centre. Furthermore, its current programming perspective is short-term in nature: locating jobs, networking, etc. However, OTI leaders are interested in moving toward intermediate to long-term goals, such as building jobs and enterprises through pro-active contacts with businesses and encouraging the development of business centres. In general, the Initiative seeks to help Ottawa's tech sector build jobs from within rather than attracting outside companies to relocate to Ottawa. One short-term gap that needs to be addressed is acquiring reliable information on what business and labour-market areas are growing and offer opportunities. There is also no organisation that maintains current, accurate and 'granular' (i.e., sub-sector-and-firm-level) regional statistics on the sector. While OTI itself may not be in a position to acquire such statistics, it certainly could encourage other relevant organisations, such as the Social Planning Council or OCRI, to do so.

Another important challenge that OTI faces is that of trying to act as a representative body for laid-off tech workers who are not only scattered across Ottawa, but who are also very heterogeneous in terms of the particular sub-sectors in which they previously worked (e.g., computer software versus telecom switching equipment) and the particular roles they have played in their companies (for example, assembly-line workers versus specialised engineers). Still, the peer-networking groups are especially supportive of OTI. The groups that served solely as 'venting' instruments for disgruntled workers did not last very long. The 20 groups that are associated with OTI seek practical results for their members. OTI is important to them.

Outside Support

To date, the most responsive outside actor has been the Ontario government, urged on by the support of the Ontario Liberal caucus. It probably doesn't hurt that the current Premier of Ontario, a Liberal, is from Ottawa and well-known to some OTI volunteers. For its part, the City of Ottawa does not have the economic resources to contribute to OTI's ongoing operations, especially in light of recent budget cuts to its own social programmes, although it did manage to provide matching funding for some additional money from the Province. And, since its own business model relies on grant-funding of projects, OCRI cannot be a financial contributor to OTI's work. So far, the federal government has not indicated substantial interest in the initiative, though OTI has some champions in Parliament. This has led to some positive meetings with staff from Human Resources Development Canada (HRSDC), for which OTI has prepared a revised proposal for additional funding. In the meantime, through existing programmes, the federal government funds an individual associated with the YMCA, who is specifically trained to deal with technology workers and now works out of the Action Centre.

Coping With Job Loss
The Special Needs of Older Workers

Research on downsizing in other sectors underscores the brutal fact that layoffs exact a heavy price from individual workers. Unemployment can negatively affect a worker's psychological and physical health. Job loss strains marriages and can trigger family conflict, even abuse. Laid-off employees lose the companionship of the workplace; and their social networks shrink. Middle-aged, mid-career workers suffer severely. Deeply rooted in their communities, they are less willing than their younger counterparts to look for work elsewhere. Moreover, often with a spouse and children to support, these older workers are also less able to afford to return to school and often must carry a heavily mortgaged home in a stagnant real estate market. In addition, the worker's spouse may still have a paying job and to relocate would mean to lose that job (Leana and Feldman, 1992).

To cope with job loss, unemployed workers should turn to government and community agencies for skill-building in job-searches, resume-preparation and interviewing. They should be able to access retraining and/or relocation programmes. Unemployed tech workers should reach out to family and other social support systems. And, they may need financial assistance to 'bridge' the period between their layoff and their next job. Moreover, company-sponsored programmes designed to ease the transition process out of one job and into another should be available as well (Leana and Feldman, 1992).

In the areas of counselling and retraining, the Ottawa experience suggests that traditional services may not meet the particular needs of tech workers. For one thing, existing programmes were designed for people without the high level of education that characterises tech professionals. This is an important rationale for OTI-provided customised services targeted to this specific constituency. On the issue of financial bridging, many Ottawa tech workers received adequate buyout packages. They then were eligible to receive Employment Insurance (EI). However, EI only lasts for one year, and many workers have now been unemployed for two to three years. Clearly, this arrangement falls short of real livelihood needs in the tech cluster.

In Seattle, after the bubble burst there, laid-off tech workers tried to move into the burgeoning biotechnology industry. Management employees with highly transferable skills were able to make this transition fairly easily. However, engineers, programmers and technical specialists, who make up the bulk of tech workers, have found few markets for their niche-specific skills. Furthermore, Seattle tech workers were as old as workers in other sectors, and as equally likely to be heads of households. Thus, they faced a number of serious barriers to coping with job loss. At the same time, what was left of the technology industry moved to an even greater reliance than before on contingent forms of labour – based on short-term, non-binding work contracts – at lower compensation levels (Doussard and Mastracci, 2003).

OTI research indicates the challenges are considerable in Ottawa as well. A survey by the Ottawa Talent Forum of 241 unemployed technology workers in early 2004 found that more than 40% of respondents were over the age of 45, and 65% were over the age of 40. Many in this group had worked for the same firm for ten to 15 years. As elsewhere, these mid-career, middle-aged workers face many barriers to coping successfully with job loss. Further, the survey found that 60% of respondents were using Employment Insurance benefits; some had exhausted their EI benefits. In addition, 156 respondents reported having gained an undergraduate university degree, and 56 had master's degrees. At the time of the survey, 80% of respondents indicated that they planned to seek employment outside their existing field (OTI Community Action Plan, 2004).

Other OTI research summarised the challenges and barriers facing unemployed, mainly older, technology workers as follows:

- Exhaustion of Employment Insurance benefits;
- French-language requirements for federal-government employment;
- Security-clearance requirements for federal-government employment;
- Lack of availability of venture capital for start-ups;
- Outsourcing of technology jobs by major companies;
- Susceptibility to exchange-rate fluctuations by export-oriented tech companies;
- Reduction of out-placement services due to the protracted nature of the downturn;
- Reluctance of non-tech employers to hire tech workers for fear that they may leave once the tech sector rebounds;
- Negative image of technology workers as over-paid and demanding;
- Age discrimination against hiring older workers;
- Lack of awareness by employers in transferability of specific tech skills;
- Lack of ability of tech workers themselves to transfer their skills to new employers;
- Obsolescence of specialised skills;
- Lack of financial assistance for training in new skills; and,
- Lack of detailed knowledge on growth trends within and outside the technology cluster.

Overall, the Ottawa Talent Initiative is well-positioned to find ways of addressing these and other barriers, and has begun to do so.

Growing the Pie: Toward Community-Based Cluster Renewal

As necessary and promising as OTI is, however, it is insufficient as an approach to tech-sector adjustment. For one thing, the Initiative's work on human resources has been constrained by money and scope. For another, working on the 'supply-side' of the labour market is only part of what needs to be done. Serious efforts, at the same time, must be made to address the 'demand-side' of the equation, through active employment and enterprise development, which OTI and other organisations are now only beginning to address. Ottawa leaders had hoped the market would naturally, fluidly, generate new jobs and businesses. It hasn't. It is time to consider a new model of cluster renewal.

Research by the Conference Board of Canada (2004) suggests that community-based organisations – such as cluster councils, networking organisations and economic development agencies – are especially important at the early stage of a cluster's lifecycle and later, at the renewal-or-decline stage. Between these points, during the growth and maturation stages of a cluster, the region's anchor technology firms are outward-oriented and well-resourced, and may not perceive the worth of such community instruments. However, when markets shift to new products or new sites, the cluster sits "on a precipice of sorts between reinvention and dissolution." Renewal strategies can include, among others, stimulating emerging technological areas that can build on the capabilities of the cluster, licensing new technologies, and forming alliances with other clusters.

In California, Joint Venture Silicon Valley is a 15-year-old non-profit organisation that has enabled that region to formulate effective responses during several periods of decline. The catalyst for Joint Venture's creation was Silicon Valley's previous recession during the late 1980s and early 1990s. Now focussing on rebuilding the regional economy in the wake of the tech bubble bursting, the organisation has taken on a wide range of policy issues – among them regulation, health care and tax policy – and has also secured significant federal and state funding to stimulate tech innovation and sector reconstruction (Conference Board of Canada, 2004).

Prior to the downturn, The Ottawa Partnership (TOP), under the auspices of the City of Ottawa, had commissioned a cluster mapping exercise to determine the breadth of the region's high-tech activity. Although there was activity in several high-tech sub-sectors, such as software and biotech, the region was historically reliant on telecommunications, the demand for which had triggered an unprecedented level of growth. Given the tendency toward high-tech specialisation in a region, when trying to develop strategies to renew Ottawa's high-tech economy, regional leaders have to avoid the temptation to adopt a generic strategy. Instead, they must closely examine what Ottawa's existing comparative strengths are and in what technology and market niches the newer local competencies are emerging. It is difficult for Ottawa to 'jump' into a completely new high-technology area, given constant changes in technology and market demands and given that past success in one high-tech field does not guarantee success in another. Moreover, developing new economic prospects in the region requires more 'indirect animation' and social capital building efforts on the part of public authorities than they are used to, but which were important for the initial rise of the Ottawa tech economy.

European experience is instructive in this regard. The European Union's Regional Innovation and Technology Transfer Strategies and Infrastructures Initiative engages targeted regions in a multi-stepped planning process. The first step is to set up a Steering Committee with balanced representation from such stakeholders as politicians, development agencies, entrepreneurs and other business interests, and providers of innovation support services, such as universities and technology transfer centres. An analysis is undertaken to assess the strengths, weaknesses, opportunities and threats (SWOT) in the regional economy. Special attention in the SWOT analysis is devoted to the comparative advantages of the region under study in relation to competitor regions. A strategic framework and specific projects to promote innovation and tech transfer are then formulated, implemented and monitored. In the political culture of Europe, this kind of multi-stakeholder process, facilitated by the state, attracts considerable support.

These approaches offer more promise than Ottawa's current approach to cluster renewal. But even the California and European models are not comprehensive enough. First, they tend to maintain the technology sector as a 'gated community' in labour market terms. Second, their key structures do not include substantial representation from unemployed technology workers. Third, there are few links with the social sector of the region or community. Social agencies pay a large price when tech clusters decline or fail outright; these agencies should be 'at the table' for that reason alone. Fourth, their enterprise development work is still indirect, attempting to stimulate the commercialisation of R&D and creating the conditions for business to flourish. Again, this is not enough.

In fact, we need to turn to the field of community economic development in order to fashion a truly broader strategy for cluster renewal. Community economic development, or CED, is a field of practice in which non-profit corporations, associations or cooperatives "blend business and social techniques to reduce poverty and unemployment and revitalize communities" – either geographic communities or communities of interest (Jackson, 2004, p. 68). In general, "the goal of CED is to promote social well-being through the creation of local economic opportunity" (Torjman and Leviten-Reid, 2003, p. 68). In Québec, the term more commonly used to identify non-state, non-market organisations active in this field is *économie sociale*.

In 2004, the Government of Canada's Department of Finance (2004, p. 16) announced a 132 million Cdn$ initiative to finance and build the capacity of social enterprises and CED organisations across the country, recognising the importance of CED strategies "involving citizens, governments, the voluntary sector, business, learning institutions and other partners." Such multi-sector partnerships share risks as well as benefits among stakeholders (Levine, Torjman and Born, 2002). This approach is most effective when CED organisations take an asset-based rather than a needs-based approach to local development, helping communities to "assemble their strengths into new combinations, new structures of opportunity, new sources of income and control, and new possibilities for production" (Kretzmann and McKnight, 1993, cited in Mathie and Cunningham, 2002, p. 2).[4]

Levine, Torjman and Born (2002) set out three basic CED models that are relevant to Canadian cities. One model is creating community wealth by supporting the development of community – or cooperatively – owned and non-profit small businesses together with community business networks. Organising community or cooperative enterprises in the high-tech sector may not be such a far-fetched idea. In fact, it should probably be considered an essential tool in any cluster renewal 'tool-kit.' The second model is community mobilisation, through organising, animating, listening, group work, strategic planning, conflict resolution, leadership development and, in particular, inclusive engagement of minorities or other citizens who may in some way be marginalised. The third model is self-employment development and micro-lending, providing customised training, business advice, mentoring, and micro-loans to enable unemployed workers or welfare recipients to start their own businesses and succeed. To complement and promote these CED models, municipalities can play several important roles, including the following: convening diverse sectors through multi-sector leadership roundtables; creating favourable conditions such as tax breaks to local firms to hire long-term, unemployed workers; employment development by supporting innovative training programmes and community loan funds; and developing and maintaining an information base, including detailed labour-force data, analysis of untapped capacity, inventory of job vacancies, available training programmes and research and methods

[4] Asset-based community development (ABCD) has become a well-known stream of community development and CED. American John McKnight has been a leader in promoting, demonstrating and assessing ABCD efforts in Chicago and elsewhere. The Community Foundation of Ottawa has invited Dr. McKnight to speak in Ottawa and has used the ABCD framework in some of its internal planning.

used by other municipalities to renew local economies (Levine, Torjman and Born, 2002).

Combining a cluster-based strategy with a CED-based strategy involves 'stretching' CED to encompass the needs of high-tech small and medium-sized enterprises (SMEs) and unemployed technology workers. Similar to cluster-based policies, CED focuses on long-term enterprise-based strategies that recognise and build on existing resources and talents in the community. Both cluster and CED strategies also recognise the importance of networking and 'social capital' building as catalytic activities. For a smaller city like Ottawa, however, it makes sense to link both strategies together. As a way to encourage economic development more generally, a key objective for all levels of government has been to create a positive environment for SMEs, with special focus on those businesses based on new products and technologies and driven by innovative entrepreneurship.

The region is increasingly made up of a proliferation of high-tech SMEs as a response to the jobs-shedding by larger firms, with these SMEs struggling to achieve sustainability and the next level of growth. There is also a new group of 'disadvantaged' citizens in Ottawa – the longer-term unemployed tech workers who are experiencing problems re-entering this sector or finding employment in alternative sectors. CED initiatives are multi-faceted, however, and deploy a range of strategies aimed at encouraging regional business growth such as training and job placement, technical assistance for business start-ups, expanding local access to capital, and job creation and retention. These initiatives have traditionally targeted the disadvantaged and long-term unemployed; but, there exists a suite of techniques and practices that can be adapted to the dynamics of Ottawa's high-technology cluster and the region's local business and labour market realities.

The Ottawa Knowledge Initiative
A Better Way Forward

There is a better way forward for Ottawa. We propose extending, expanding and converting the Ottawa Talent Initiative into a community-based approach to cluster renewal – we call it the Ottawa Knowledge Initiative (OKI). The OKI would comprise six components: community-based organisation, enterprise generation, a social-tech incubator, enterprise financing, community workforce intermediaries, and database management.

Community-Based Organisation

A non-profit legal structure would be put in place to guide OKI, with substantive representation on its board of directors not only from unemployed tech workers and executives of technology companies, but also from leaders of social organisations, the community (foundations, major non-profits), universities and colleges, and governments at all levels. In particular, the OKI organisation would establish a multi-sector leadership roundtable to develop a detailed plan for community-based cluster renewal – one that would meaningfully engage all of Ottawa's diverse sectors and constituencies.

One interesting model is Waterloo's Opportunities (OP) 2000 leadership roundtable that included not only business but the social sector, unemployed citizens and the regional government. The roundtable strove to reduce poverty and unemployment in Waterloo not only via its own initiatives, but also by each representative within his or her own sector. For example, business representatives involved in OP 2000 encouraged their peers to examine their employment practices around hiring, wages, working time, and layoffs (Levine, Torjman and Born, 2002). Another model is the Joint Venture Silicon Valley organisation, which has drawn attention as an example where a local community employed specific strategies to changes its economic trajectory. Joint Venture brings together established and emerging leaders form the business, labour, government, nonprofits, etc. to create an *economic community* – that is, where there is a strong, responsive relationship between the economy and community that allows for sustained advantages to accrue both to companies and to the community.

In terms of practical next steps, OTI should become a separate legal entity. Its mandate, board and programme capacity should all be expanded to take on the role of catalyst and driver of community-based cluster renewal. It should be rebranded as the Ottawa Knowledge Initiative. And, substantial funding from all levels of government should be made available to support OKI's activities. In some areas, OKI could sub-contract with existing local organisations to provide certain services. For example, OCRI could be tasked to promote tech start-ups on the supply side, while OTI builds up its services on the labour-market side. The Social Planning Council could assist in database management or other functions. OKI would thus lead cluster renewal but not necessarily directly deliver all the essential services.

Enterprise Generation

Public and charitable funds would be mobilised by OKI to support the start-up and growth of technology, and technology-related, small enterprise. Cooperative and non-profit business forms would be encouraged as much as privately-owned ones. OKI would provide assistance for business plans, market studies, research and development, and financing proposals. Small business could also access OKI funds to hire consultants on business strategy and management, human resources and other aspects of enterprise implementation. Furthermore, OKI funds would be used to promote exports to the US, Europe and Asia. Where a product or service matched a market, and no individual entrepreneurs were found to come forward, OKI itself would *directly* create the enterprise and grow the business with the intent of spinning it off to other owners later.

Researchers at the University of Ottawa have also emphasised the need for a proper technology incubator in Ottawa. But, they eschew the standard spin-off model where it is assumed that technology in an incubator has to be commercialised almost exclusively as a new product in a spin-off company. Not all technologies can provide the basis for a company and establishing a company is a very onerous process unto itself. Instead, the role of an incubator should be to assess the commercial potential of a technology, in conjunction with the market conditions and the venture's management team. In addition, at a critical juncture, an assessment needs to be made whether the technology should be spun off to a start-up or licensed or sold to a large company. It is more likely that technologies, rather than companies, are to be brought to market. The sector may be able to achieve more modest and sustainable successes this way. This would be a better foundation for regional growth rather than encouraging the inflated expectations of would-be entrepreneurs (Large, 2005). This would seem to be wise counsel in the case of Ottawa.

Social-Tech Incubator

Another instrument that could blend the social enterprise and CED strategies is a *social technology* incubator. This would be a community-owned corporation that would support entrepreneurial individuals and groups in R&D, business planning, marketing, financing and licensing *to create and grow businesses that use technology to meet social needs.* There are many applications of software, hardware, telecom, photonics and biotechnology that can, among many other things, help fight poverty, improve the environment and energy efficiency, promote human rights, and meet the needs of persons with different abilities. The social technology incubator would not only seek to match technology innovations with social organisations and constituencies, it would also put together the money, expertise and markets to bring the technology application to commercial reality – for a social purpose.

One of the most interesting models for a social tech incubator is the Palo Alto-based Beneficent Technologies, or Benetech. The creation of former rocket scientist and tech entrepreneur Jim Fruchterman, Benetech is actually three organisations – two non-profits and a for-profit – that share leadership, space and services. An overall board of directors provides senior-level expertise in intellectual property, venture capital, software, life sciences and philanthropy. Benetech plays the role of in-house venture capitalist, evaluating product ideas, supporting market research, creating product 'buzz,' and mobilising financing – especially foundation and corporate grants. It organises enterprise teams, and supports them in preparing rigorous, benchmarked business plans that aim to quickly achieve break-even financial performance and high social impacts with a longer-term self-sufficiency or exit plan.

Two projects illustrate Benetech's approach and potential. Bookshare.org is an online library of 10,000 scanned, downloadable books that can be accessed in audio or Braille for a small fee by blind or reading-impaired customers. This social enterprise is growing through partnerships with publishers, software companies and post-secondary institutions. The second project is Martus, which provides human rights workers with web-enabled software tools to collect analyse and disseminate information on human rights abuses. Using open source technology, Martus generates revenue from server maintenance and administration, customising software and user training. Other Benetech projects involve testing technologies for large-scale statistical studies in the human rights field, landmine detection, handheld computers and camera systems for the mobility-impaired community, and open-source software to increase computer use among low-income groups.

While none of these businesses is likely to create thousands, or even hundreds of jobs, such social enterprises do create some jobs – along with a lot of positive momentum in the community. They give confidence to entrepreneurs, both individual and collective, and engage key actors such as foundations and social activists. There are obvious collateral benefits associated with bringing these parties together with the tech community to work on joint projects. The bottom line here though is to ensure that these businesses do, in fact, meet important social needs. Their *social* return on investment can be very significant, indeed.

Enterprise Financing

OKI would also inject new capital into the non-profit Ottawa Community Loan Fund, and strengthen its ties to OCLF. The Fund would become the Initiative's financing arm to enable new enterprises to access appropriately targeted and packaged low-cost loans and guarantees. OCLF would, in turn, bring other financial institutions – including the Business Development Bank of Canada, banks and credit unions/*caisses populaires* – to the table to share the benefits and risks of lending to new tech start-ups.

Community-Workforce Intermediation

OTI has played a useful role in assisting unemployed workers to tap into a wide network of contacts through which job seekers can learn about potential job openings and share experiences. Intermediaries are important in this regard as they operate as the 'linking agents' between the needs of job seekers and employers. However, community organisations in Ottawa that are positioned to play such a role often lack the sectoral expertise to be fully effective – that is, they are staffed by professionals with social service backgrounds but who are unfamiliar with the occupational capacities and circumstances of technology workers. An OKI-type organisation could more fully serve as a workforce intermediary between the high-tech sector and unemployed technology workers. One important experience to draw from, for example, is that of the Learning Enrichment Foundation (LEF) in Toronto. LEF has established close working relationships with a number of sectors in Toronto. Based on its intimate understanding of the skills needed by local businesses, as well as the skills and capacities of the long-term unemployed workers who are their clients, the Foundation has developed a suite of programmes to help transition these individuals into these sectors. In particular, LEF pioneered a 'Life Learning System' to formalise its targeted training efforts and give it the ability to link these efforts with small and medium-size companies (Torjman and Leviten-Reid, 2003).

Database Management

Another important component of OKI's work would be to maintain a comprehensive, local-level database – derived from a one-time household census of employment status and skills and knowledge assets, updated by smaller-sample studies on a regular basis. This information would be used by OKI to target, adjust or request certain training offerings for unemployed tech workers. The Initiative would also link the talent in the database with the immediate labour-market needs of small enterprises and large corporations alike – that is, matching the supply of skills with demand.

The census would involve a massive training and implementation effort for the largest household survey ever carried out in the Ottawa area. The management information system to maintain and use the collected data would require professional management. With its labour-market initiative 'TalentWorks,' OCRI already tracks changes in labour supply and demand in the region, but this could be deepened to have a more sub-sectoral focus. In addition, a related tracking system would involve a community-based 'early warning' systems that could help identify larger companies that may be in trouble. The CED corporation RESO in Montréal developed a capacity with local stakeholders to detect firms in trouble and, through a broad-based community process, was able to reduce the outward exit of businesses and retain the jobs at risk (Levine, Torjman and Born, 2002).

Conclusions

Steering on black ice is never easy. It really is only the possibility of community solidarity and assistance that makes trying to steer worthwhile at all. Ottawa is trying hard to steer on the black ice of a stagnant, unpredictable tech sector. In such an environment, its best option is to adopt a community-based approach to cluster renewal. The region has begun to take some initial steps in this direction, but there is much more work to be done. A clear vision, strong political will and adequate public funding are the keys to success. In the meantime, whether the Ottawa tech cluster can really deliver sustainable livelihoods to large numbers of workers and citizens in the region still remains a serious and open question.

Epilogue

In mid-2005, the authors circulated a draft of this chapter to Ottawa stakeholders for their comments. With few independent analyses of the region's tech sector available in the public domain, the media quickly picked up on our critique of the sector's shortcomings, triggering a nearly year-long series of interviews and stories. Like it or not, we became *engaged scholars* – direct actors in the debate we had been observing from the sidelines.

First, local radio and television carried stories of our account of the impact of tech's downturn on workers and families. Our criticism of OCRI's expanded definition of sector employment was highlighted.

Tech-sector boosters were not amused, and took us to task in the media and at conferences. By late 2005, the sector was, in fact, showing signs of a revival, and its leaders were not interested in more sober observations about the sector's workforce realities. We decided to soften our stance on the jobs-definition issue, agreeing that what really matters are new-job creation, sustainability and quality – regardless of what baseline employment estimates are used. For its part, OCRI began citing the job-growth data of Statistics Canada. An uneasy truce was reached on this matter.

Early in 2006, however, the sector's concern had shifted towards the question of the supply of skilled labour. Technologies and business models were evolving rapidly, requiring very specialised and often novel skills and knowledge. Noting that university enrolment in computer engineering and computer science had declined steadily in recent years, OCRI's President emphasised that "we need to make sure that highly qualified workforce personnel are available to fill the increasing demand" (OCRI, 2006, p. 2).

How should this be done? Some CEOs and their advisors called on regulators to give them more flexibility – through more contingent (short-term contract) workers, targeted immigration and off-shoring of jobs – to be able to recruit employees with the new skill sets needed to maintain growth. Not surprisingly, we countered this approach by arguing that there was sufficient talent that could be redeveloped in the region already, and that more contingent work, whose benefits are minimal and make stability risky, would be bad for families and the community. We also questioned the quality and stability of employment in the many start-ups and small businesses now accounting for much of tech's current round of growth (Pappone, 2006a).

By mid-2006, a year after our chapter had 'gone public,' the Ottawa Talent Initiative had gained federal funding and was playing a more active role in animating this public debate. It was also providing a somewhat wider range of services to out-of-work knowledge workers in the region. In a major conference, *Tech Trends: 2006 and Beyond*, OTI focused on bringing the key stakeholders together in order to once again assess the state of Ottawa's technology sector and to construct a new plan for the future. As OTI's Executive Director put it: "The number one thing that people want to know is what is happening in the sector and where is it going, and that's where we are finally trying to get" (Pappone, 2006b, p. 11).

We believe that our recommendations remain key to the region's ability to navigate toward sustainable livelihoods in the technology sector. Broadening OTI's community representation, encouraging it to promote enterprise generation and financing and database management, and experimenting with a social-tech incubator – these actions could define an Ottawa *Knowledge* Initiative that would advance the interests of technology workers, their families and the community at large in the challenging years ahead.

References

Bagnell, J. (2005a). $500 billion later. *Ottawa Citizen*, March 3, F4-F5.

Bagnell, J. (2005b). Few startups standing after the party ended. *Ottawa Citizen*, March 3, F5.

Beckstead, D., Brown, W. (2005). *An anatomy of growth and decline: high-tech industries through the boom and bust years, 1997-2003*. Ottawa: Statistics Canada.

Buss, T.F. (1993). A tale of two cities: revitalizing Corby, Northamptonshire, and Youngstown, Ohio. In P.B. Meyer (ed.). *Comparative Studies in Local Economic Development*. London, England: Greenwood Press.

Conference Board of Canada. (2004). *Clusters of opportunity, clusters of risk*. Ottawa: Conference Board of Canada.

Cortright, J., Mayer, H. (2002). *High tech specialization: a comparison of high technology centers*. Washington, DC: The Brookings Institution.

Department of Finance. (2004). *A new deal for communities: budget 2004*. Ottawa: Department of Finance.

Doussard, M., Mastracci, S. (2003). *Uncertain futures: the real impact of the high-tech boom and bust on Seattle's IT workers*. Chicago: Center for Urban Economic Development, University of Illinois.

Feldman, L.C., Feldman, D.C. (1992). *Coping with job loss*. New York: Lexington Books.

Godinez, V. (2004). Experts debate future demand for IT workers. *Ottawa Citizen*, January 10. K4.

Goff, K. (2003). Few pay gains for skilled tech workers. *Ottawa Citizen*, November 18. E3.

Goff. K. (2004a). Office vacancies near 7-year high. *Ottawa Citizen*, September 25. D1.

Goff. K. (2004b). Full-time employment slides again. *Ottawa Citizen*, December 4. D1, D4.

Goff, K. (2004c). Funloving Ottawa: we're well read, so-so with attire, and highly taxed. *Ottawa Citizen*, December 14. D1, D4.

Goff, K. (2005a). Ottawa: the year ahead. *Ottawa Citizen*, January 4. D1, D4.

Goff, K. (2005b). Shifting gears: consumers, retail take lead. *Ottawa Citizen*, January 5. D1, D4.

Goff, K. (2005c). Growth in Ottawa home prices slows in final quarter. *Ottawa Citizen*, January 8. D1, D4.

Goff, K. (2005d). Technology sector adds 3,700 jobs in January. *Ottawa Citizen*, February 5. D1.

Goff, K. (2005e). Getting back to the growth groove. *Ottawa Citizen*, January 26. C1, C4.

Hill, B. (2004a). Why the shine is off semi-conductor sales. *Ottawa Citizen*, October 8. E1-E2.

Hill, B. (2004b). Venture capital funding in Ottawa jumps by 33% in third quarter. *Ottawa Citizen*, October 23. B1.

Hill, B. (2004c). Zarlink says high dollar will double loss. *Ottawa Citizen*, December 8. D1, D4.

Hill, B. (2004d). Looking back at the year in tech. *Ottawa Citizen*, December 30. E1-E3.

Hill, B. (2005). Nortel to shut Carling campus research labs. *Ottawa Citizen*, January 15. D1.

Hum, P. (2005). Back from the brink. *Ottawa Citizen*, March 10. E1, E4-E5.

Industry Canada. (2004). *Information and communications technologies statistical overview*. November. Ottawa: Industry Canada.

Jackson, E.T., Khan, R. (2003a). *Sustaining livelihoods in the new economy: constructing a role for community economic development in technology cluster growth*. Ottawa: Caledon Institute of Social Policy.

Jackson, E., Khan, R. (2003b). Five ways to help former tech workers. *Ottawa Citizen*, December 19. D6.

Jackson, E.T. (2004). Community innovation through entrepreneurship: grantmaking in Canadian community economic development. *Journal of the Community Development Society*, *35* (1), 65-81.

Jarboe, K.P. (2000). *Community workforce partnerships*. Washington, DC: Athena Alliance.

Kristal-Schroder, C. (2004). Food bank struggles to meet demand. *Ottawa Citizen*, October 7. B2.

Large, D. (2005). The renaissance of Ottawa's ICT community through joint initiatives. *The Journal of Public Sector Management*, *35* (1), 1-8.

Levine, B., Torjman, S., Born, P. (2002). Community economic development in Canadian cities: from experiment to mainstream. In C. Andrew, K.A. Graham and S.D. Phillips (eds.). *Urban Affairs: Back on the Policy Agenda*, Montréal/Kingston: McGill-Queen's University Press, 201-220.

Mallett, J.G. (2002). *Silicon valley north: the formation of the Ottawa innovation cluster*. Ottawa: Information Technology Association.

Mayeda, A. (2004a). Job growth falters in capital region. *Ottawa Citizen*, September 11. D1.

Mayeda, A. (2004b). Ottawa technology hall of fame. *Ottawa Citizen*, November 24. D1, D3.

Mayeda, A. (2004c). Breathing new life into life sciences. *Ottawa Citizen*, November 27. D1, D4.

Mayeda, A. (2005). Seeking a bird in the hand. *Ottawa Citizen*, March 2. C1-C2.

O'Neill, J. (2003a). Techies on the edge. *Ottawa Citizen*, December 12. F1-F3.

O'Neill, J. (2003b). Disillusioned workers urged to stay in Ottawa. *Ottawa Citizen*, December 13. D1-D2.

O'Neill, J. (2004). Techies embark on journey out of 'black ice of life.' *Ottawa Citizen*, January 17. D1-D2.

O'Neill, J. (2005). Failure to combat child poverty a 'national disgrace.' *Ottawa Citizen*, January 22. E1, E3.

Ottawa Talent Initiative Project Team. (May, 2004). *OTI community action plan: Ottawa*. Retrieved from http://www.ocri.ca/about/assets/OTI_CAP_Final_May04.pdf).

Ottawa Talent Initiative. (2006). About OTI. Ottawa Talent Initiative. Retrieved on July 14, 2006 from http://www.ottawatalentinitiative.ca.

Pappone, J. (2006a). Labour challenges demand team effort from all players. *Ottawa Business Journal*. February 10, 2006, 1, 3.

Pappone, J. (2006b). Tech forum, job fair wake-up calls for region's talent-hungry companies. *Ottawa Business Journal*. June 26. 11.

Radko, R., McQuaid, J., Barclay, B., Putinski, P. (2004). Focus group interview with Ottawa Talent Initiative task force leaders. Conducted by Rahil Khan, Ottawa.

Savoie, D.J. (2000). *Community economic development in Atlantic Canada: false hope or panacea?* Moncton: Canadian Institute for Research on Regional Development.

Srivastava, S., Theodore, N. (2004). *America's high tech bust: a report to the Washington Alliance of Technology Workers*. Chicago: Center for Urban Economic Development, University of Illinois.

Sutcliffe, M. (2004). Ten years after: back to potential? *Ottawa Citizen*, September 25. D1-D2.

Sutcliffe, M. (2005). 100,000 tech jobs by '08: OCRI. *Ottawa Citizen*, January 22. D1, D4.

Tödtling, F., Trippl, M. (2004). Like phoenix from the ashes? The renewal of clusters in old industrial areas. *Regional Studies*, *41* (5/6), 175-195.

Torjman, S. (1999). *Reintegrating the unemployed through customized training*. Ottawa: Caledon Institute of Social Policy.

Torjman, S., Leviten-Reid, E. (2003). *Innovation and poverty reduction*. Ottawa: Caledon Institute of Social Policy.

Woolhouse, K. (2004). 2004 gave Ottawa-focused investors a rough ride. *Ottawa Citizen*, December 30. B1, B4.

CHAPTER 6

The Renaissance of Ottawa's ICT Community Through Joint Initiatives

David LARGE

School of Management, University of Ottawa

Introduction

The Ottawa Information and Communications Technology (ICT) community has gone through some rough times in the last four years, but now seems to be on the rebound. However, there have been some painful lessons learned, and market realities that may not have changed that command remedial attention. Government, business and civic sector groups must work together to ensure that the ICT renaissance is accompanied by new joint initiatives to help ensure even greater strength and longevity. This chapter suggests a few possible directions for action.

In the last five years, there has been considerable pain in the Ottawa region's information and communications technology sector that goes well beyond the large and visible layoffs from the major companies such as Nortel and Alcatel. In particular, there has been a string of closures and US buyouts and/or US relocations of many, if not most, of our most promising locally-grown companies. For example, there have been closures of such well-funded start-ups as Zenastra, BitFlash, Ceyba and Innovance; the takeovers of Newbridge by Alcatel, Catena and Akara by Ciena, Dy-4 by Curtiss-Wright, Corel by Vector, and Accelio (originally Jetform) by Adobe; and the 'consolidation' of the JDS Uniphase headquarters to San José. These latter occurrences have substantially compromised local companies' independence, and diminished the local marketing and production executive expertise and muscle.

The good news is there have been recent indications of the ICT sector's renaissance. This includes a recent optimistic prediction from

Jeffrey Dale, the President of the Ottawa Centre for Research and Innovation (OCRI) that the sector's local employment will increase from its current level of 67,800 to 100,000 within three years (see Sutcliffe, 2005). However, I believe that there is much to be learned from the recent period of pain, and that there are some key market realities that may not have changed and that command remedial attention. In this article I'd like to offer my personal views on some possible joint initiatives that could be undertaken by business, government and civic sector groups to strengthen the recovery and increase the longevity of home-grown companies.

Before I present my views, I'd like to discuss what I feel are some key market forces at work in the ICT industry that are largely beyond our control, and that may not have changed over the last five years. We need to recognise these external constraints if we aspire to create prosperous, long-lived and independent ICT companies. In my view, the primary key force is the power of the US market; the second is the need to be, or to appear to be, a US company in order to win in the US business-to-business market; and the third is the need for a supplier to be perceived by the customer as relatively big versus the supplier's competition.

The Power of the US Market

To be a stand-alone long-lived North American ICT company at the top of one of the value chains, a company needs to be successful in the US market, which currently has about 40-50% of the world market for just about any ICT product category. Even tougher, a large percentage of the ultimate buying power is concentrated in just a few large organisations (e.g., the US government, a handful of the major telecom carriers and financial institutions, and the other members of the US Fortune 1000), so breaking in is extraordinarily difficult. Yes, a Canadian company can get started in the Canadian market, but moving past the survival and early growth stages of a company to a sales level approaching 50 million Cdn$ requires a presence in a very tough-to-crack US customer base. This observation by itself is not news to any ICT company president in Ottawa, but it does lead directly to my next point, which many find contentious.

The Need to Be Perceived as a US Company

In his August 22, 2003 article, James Bagnall (p. A1) gives four principal reasons for constant flow of ownership and/or headquarters to the US (i.e., relative Canadian weakness in marketing, control of the start-ups by US venture capitalists, US government procurement processes favouring small US businesses, and US customers' tendencies to 'buy American').

Chief among these (in my view) are the US government's tendency and US business customers' tendencies to 'buy American.' This preference is rarely overtly mentioned by buying organisations, but I suggest that it is an unwritten consideration in every major purchase by a US organisation, whether public or private. In particular, when a purchase is advanced technology and mission or system critical (such as ICT), and especially if the customer is the US government, it could be a point of patriotic pride that the supplier be American. And now more than ever, if the technology is related to domestic security and/or intelligence services, there may be even more pressure to buy from American suppliers to keep the technology 'at home.' This tendency has in fact been made vividly clear several times in the last few years. For example, after September 11, 2001, only US companies were allowed to access certain US government ICT contracts (Tsaprailis, 2004, p. 9); in 2003, there were US proposals to end Canada's status as a non-foreign defence supplier (Greenaway, 2003, p. A3); and following the Iraqi war, Canadian companies were barred from bidding on US-funded reconstruction projects (Alberts, 2003, p. A9).

To circumvent this tendency of US organisations to 'buy American,' many Canadian ICT companies decide to become a US company outright (as Akara did), or to take on the appearance of a US company, or to disguise their true country of origin. As James Bagnall points out, it may be to dodge this 'buy American' barrier that Entrust established its headquarters in Dallas, a tactic that may indeed have led to the US State Department's decision to standardise on Entrust encryption software in 2001. It's the same reasoning that has convinced an entrepreneurial colleague of mine, a manufacturer of advanced ceramic armour, to move his head office and manufacturing operation to upstate New York from Ontario. From conversations with armour buyers in the US government, he inferred that his odds of success would be much greater if he was headquartered in the US

The Need to Be Perceived as the Largest Competitor

In *Crossing the Chasm* and *Inside the Tornado*, Geoffrey Moore emphasises his view that high-tech winners must be perceived by potential customers as the largest competitor in their class. In high tech, where purchases tend to be expensive and mission-critical, conservative mainstream customers tend to prefer big suppliers that can provide total solutions, who can also avoid bankruptcy so that they'll be around to service the product and provide next generation solutions. Small ICT providers are just bad news to mainstream customers. This was underscored in another James Bagnall article concerning the demise of Catena – expected sales to the four regional Bell companies in the US failed to materialise because "there was a limit to the Bells' willingness to bet on technology from startups" (Bagnall, 2004, p. F1).

And it's not just the buyers' preferences that influence a supplier to become big. Big *competitors* are also forcing suppliers to become big. For example, James Bagnall mentioned the sweetheart deals being offered by Newbridge's largest competitors prior to the Newbridge takeover by Alcatel in 2000 (2003, p. F2), that Newbridge did not have the muscle to match. So, a company can't stay small for long and hope to be on top – there is a relentless pressure to grow big, or to get acquired by someone who is big, or to be content with being a minor niche player.

But given the parallel requirement to be seen as American, the odds of a Canadian company becoming the largest player at the top of a value chain, or a 'gorilla,' as Moore puts it, are very small. This only underlines the uncommon excellence that has been displayed by players like Nortel and Cognos in achieving their US and global leadership status.

So, We're Early to Mid-chain Forever?

It would be discouraging enough to believe that from now on, almost none of our smaller ICT companies are to be able to achieve a top-of-the-chain position because they are to be inevitably forced to move south, or to get swallowed by a US company and become marginalised.

But wait – it gets even tougher than that. Over the next 30 years, both China and India are on their way to becoming economic powerhouses, conceivably in the same order of magnitude as the US These two countries are just now finding their economic legs, and should they accelerate at the same rate that Japan did in the 1960-1990 timeframe, they are to be formidable indeed. And I don't think there's any reason to believe that they can behave any differently than the US when it comes to purchasing mission critical supplies like ICT. I think they will favour and nurture their own, leaving Canadian ICT suppliers even further on the sidelines.

So, do we throw in the towel? I say *no*. If we accept the probable long-term role of early to-mid-chain suppliers, let us then learn how to maximise the probability of success in this upstream role, and to maximise the return from each success story. If we can learn better models now in building our partnership relationships, especially with the US players, then we'll be in much better shape for developing our partnership relationships with China and India as they build momentum.

The first thing we can do, of course, is to do better with the small companies we have already because the longer a company can hang in, the better its bargaining position, and the larger the mass that might stay stationed in Ottawa. How to more effectively grow our small companies to about 50 million Cdn$ in annual revenues is a central challenge, and this issue is being currently studied in depth by OCRI.

But it is not enough, in my view, just to do more of the same, only better. Let us all try to think in new directions, to help create new models for how we can direct and nurture our ICT technologies and our ICT technology companies. Here are a few suggestions, based on observations of successful models tried elsewhere.

Nurture More Consumer-like Products in our ICT Portfolio

It appears that most of our start-up ICT companies tend to favour the development of products or components embedded in larger systems for purchase by businesses and governments, which means that the number of customers is relatively small, and they're probably biased towards purchasing from larger suppliers, and the key US buyers may be biased towards American suppliers. One way to avoid these organisational purchasing biases is to move towards (or begin with) consumer-like products (e.g., products with potentially millions of individual purchasers). US consumers have shown a much higher tendency to accept superior 'foreign' products than businesses have, witnessed in spades by their acceptance of Japanese cars and cameras in the 1970s and 1980s, all the way up to the present by their sustained acceptance of Nintendo and Sony, and products from European companies like Nokia.

Some Canadian technology companies have played the consumer-like product strategy very successfully (e.g., RIM with Blackberry, and I would argue Corel with CorelDraw, which had heavy appeal to individual graphic artists). In our community today, I think that Zim is a company to watch specifically because its short-text messaging products have a lot of potential appeal to the US consumer base. Should Zim ever achieve sales of 50 million Cdn$, I'd give it a very high probability of maintaining its independence for a long time, and perhaps sustaining a dominant position. Note that there are big, big differences involved in selling to the consumer, so a migration towards consumer products should not be undertaken lightly. Separate operating divisions, and a different skill set and mindset, are probably required.

Promote New Partnership Models, like 'Pre-emptive Joint Ventures'

One of my entrepreneurial software colleagues in Ottawa is currently in the process of finalising an agreement that I shall call a 'pre-emptive joint venture.' In this relatively underemployed model, the Canadian start-up company co-establishes a US-based joint venture (a separate, co-owned legal entity) with a US technology company, each contributing highly complementary intellectual property. The Canadian company takes a significant, but minority, ownership position, while the US company takes the majority ownership position, and thus the joint venture company remains an American company.

The JV's prospects are much better than those of a Canadian company on its own, as the JV is American, has stronger intellectual property, and has a much stronger local base of customer contacts. Thus, the Canadian company retains a large interest in an American start-up company that has a higher probability of success, gets licensing revenues for its IP, and also retains a fully independent Canadian company. It is called 'pre-emptive' because it may pre-empt the nearly inevitable closure or buyout.

Nortel has employed a very similar model for continuing its pursuit of the rapidly-emerging Chinese market. It recently announced a joint venture agreement with China Putian Corp. to develop and build cellphone equipment to Chinese standards, whereby "China Putian will own 51% of the joint venture and Nortel 49 percent" (Hill, 2005, p. E1).

Build an Independent Industry-Focused R&D Lab

There is no shame in being highly, highly competent in technology R&D. We have the R&D talent, with possibly the highest concentration of technology PhDs in the country, so let's use it! There's a model in the US that may be useful to look at – the Battelle Labs based in Columbus, Ohio. It is a non-profit lab with a strong commercial orientation that employs 16,000 and conducts $2.7 billion USD in diversified annual research for some 2,000 companies and government agencies.

I'd like to suggest that we consider building a lab something along those lines, which would offer a range of technology research services to customers worldwide (e.g., to Indian and Chinese organisations as a start). As a stretch goal, we could possibly achieve a lab with some 1,500 employees, with an annual revenue stream in the order of 400 million Cdn$. Given the number of technology PhDs currently looking for work locally, there might not be a better time to get going.

The proposed lab would not at all be intended to compete with our existing federal government labs, whose primary mandate is to research and develop technologies of national importance that are not necessarily suitable for commercialisation, and whose industry relationships tend to be restricted to Canadian companies. We need our federal labs to continue to fulfill their mandate as they are doing, while the proposed lab would cater to a much different clientele with a different set of operating objectives and constraints.

Build a 'Total Solution' Incubator

Discovering technology in an R&D facility is one thing. Moving it towards commercial adoption is very different, and in my view, technology incubators and their cousins are an essential part of the commercialisation engine we are trying to build. We do have many variants operating in the region, the Ottawa Biotech Incubation Centre being the most complete and exciting version that I've seen. But I still think we haven't quite got the total solution in place from the venture team's point of view. There seem to be three chronic limitations with incubator models (e.g., 1) an over-emphasis, in my view, on start-up companies as the optimal commercialisation mechanism; 2) a crippling absence of guaranteed *and sufficient* seed money upon admittance to an incubator; and 3) venture management teams that are incomplete and likely short on management skill and/or experience).

First, there are many paths for bringing a technology to market, not just a start-up company. For example, a basic technology can be licensed or sold outright to a large operating company, it can be used as the focal point for the formation of a stand-alone incorporated joint venture, or it can be licensed or sold to a start-up company with the 'spin-out' model. What I see in practice is a very large assumption that a technology in an incubator should be commercialised almost exclusively as a new product in a spin-out company. This can be a poor strategy, as the mechanics of establishing a company are incredibly difficult and time consuming, customers are perhaps biased against start-ups, and, frankly, not all technologies have enough legs to form the foundation of a company.

Therefore, it is a vital role of an incubator, in my view, to objectively assess the commercial potential of the technology itself in conjunction with the market conditions and the venture's management team. And, the incubator needs to assess at a critical gate whether the technology should be spun out to a start-up, or simply licensed or sold to a large company. Using this approach of different but more appropriate mechanisms, many more technologies are to see the light of day with more modest success, perhaps to the disappointment of overly optimistic entrepreneurs, but to the greater benefit of the region. There is a lot to be said for using the brand recognition and corporate infrastructure of an established entity as 'the best' exit mechanism from an incubator.

Second, incubator venture teams simply need more seed money, guaranteed up front, upon admittance. A technology has to be developed into a prototype product for use by potential customers in order to permit any valid assessment of whether there is any relative value in use, whether the necessary launch window can be attained, and what mechanism (e.g., license, JV or spin-out) is the most appropriate. To reach that stage of development is relatively inexpensive (perhaps a few hundred thousand up to a million dollars), but sufficient funding *must* be attained. And it must be *guaranteed* for a certain period of time so that the venture team can focus on technology and customer issues rather than finance issues.

It has always seemed sub-optimal to me to make start-up entrepreneurs spend most of their time hustling to find this money – it is far better that we have them focus on completing beta prototypes and getting on the plane to visit with customers rather than hunting for cash. The old model of requiring the mortgage of the family home keeps away many potentially successful entrepreneurs at a critical time in their lives, or the requirement to spend their time hustling for money simply kills the enterprise. We would all be better off to give a core venture team a very modest stipend for a guaranteed period of time, and have them focus on getting the job done.

Third, incubators need a rolling supply of battle-tested and dedicated venture management teams whose new venture skills have been intensively honed, to migrate 'up or out' with their venture teams. Such teams could be developed by a preparatory boot-camp programme with a mix of both seasoned veterans and younger 'top guns' who receive an intense classroom experience, then intern with a venture group developing a technology, and then migrate out with the venture team should it proceed to a joint venture or spin-out.

The classroom portion could be delivered by an extremely competent team pulled from the ranks of both Carleton University and the University of Ottawa (jointly, the two schools have a half-dozen of the best professors in Canada in this field) augmented with classroom participation by seasoned industry veterans. Importantly, the members of the venture management team would receive a very modest stipend from the seed capital fund while they are in the incubator (enough to pay a few of the essential bills at home).

As it is, there are thousands of extremely talented and motivated people in the region who would be willing to work for a very modest wage to be a part of an exciting new venture team, but almost none who can afford to work for free. I think that in order to progress from here, then changing the ruthless attitude that 'if I had to suffer to make it work, then everyone has to suffer to make it work.' Let's think about more effective ways to improve our success rates!

These limitations could be eliminated by a true 'total solution' incubator, and I would not be the first one to suggest this. For example, in a book written back in 1986 by Raymond Smilor called "The New Business Incubator," Ray was even then calling for a total linkage of "talent, technology, capital and know-how" under one roof. In the book, several types of incubators were discussed, but there was one in particular that seemed to have answers or near-answers for all the limitations just described. According to Smilor, The Rubicon Group Incubator provided a near-total solution in return for an equity stake in the ventures. It offered seed money for up to two years, full consulting legal and administrative services, information systems, and arguably most importantly, business education for the technological members of the team. Of its initial 10 companies, after one year, four remained very healthy, and were expected to achieve revenues of 7-10 million Cdn$ within the following five years.

As another potential benchmark, a venture capital company called Arts Alliance Ltd. may have an outstanding model to study. While not formally an incubator, the company nonetheless seems to be providing most of the full array of services advocated above. That is, Arts Alliance takes on early stage companies, and methodically mentors them over a longer period of time, rather than "flipping them" (Balu, 2000). The results are impressive. From 1997 to 2000, Arts Alliance invested in 36 companies, realising an aggregate ROI on its first six exits of 911%, with only two folded companies (Balu, 2000). This compares *very* favourably with the more typical VC results of only 2 funded ventures out of 10 providing an acceptable return upon exit through an IPO or sell-out. I do acknowledge that their impressive statistics were achieved in the over-heated run-up in the late 1990s, but the results do warrant a closer look.

Build a 'Super-Brand' to Impart Respect and Confidence

In the world of ICT, there has been a growing recognition among the major ICT suppliers over the last decade of the value of establishing a strong corporate brand that imparts feelings, among others, of respect and confidence. Companies like Microsoft, Intel, and Cisco have individually spent hundreds of millions of dollars on corporate branding campaigns to establish the desired company image, as well as to establish positive effects and confidence in the minds of potential customers. Similarly, we have witnessed the beginning of an expensive new campaign by Nortel to re-establish a positive image.

To me, it seems like a good notion to intentionally establish a technology super-brand for the region that our start-ups can be associated with, and that would convey to customers worldwide some sense of greatness of technology and product, greatness of management, and longevity of company. In other words, the brand would essentially provide a stamp of approval that would impart far-reaching respect and confidence, helping potential customers to overcome their fear and uncertainty surrounding small technology companies. Brands such as 'NASA' or 'Batelle' come close to the magnitude that would be useful. It is possible that if we did construct the 'total solution incubator' as previously discussed, the super-brand could be built around the name of that incubator, and could thus be associated with every one of its ventures, with the intent of helping the venture open doors that would otherwise not be open.

Conclusions

In light of certain market conditions in the ICT sector, I've suggested a few initiatives we could tackle jointly as a technology community, such as more consumer-like products in our product mix, 'preemptive joint ventures' with US (or Chinese) partners, an independent industry-focused R&D lab for hire, a total solution incubator that comes with a big pot of guaranteed seed money and a steady stream of finely-tuned venture managers, and a super-brand that connotes world-class technology commercialisation capability.

In my own view, I think the idea of the total solution incubator emerges as the first priority. If we learn to choose our venture candidates more wisely than other countries, give them the financial and managerial nourishment they need to quickly achieve a prototype stage, and select the optimum commercialisation mechanism, then we will have many more successes. There is one missing link – the pot of 25 million Cdn$ or so to serve as the initial source of incubator start-up funding and venture seed funding. But given the hundreds of millions in venture financing that have simply vapourised in the region in the last few years, and the hundreds of millions that the federal government spends annually on technology studies and various R&D tax incentives, surely putting up an initial pot of 25 million Cdn$ from cooperative sources to create the world's best technology commercialisation engine is a worthy undertaking.

I am personally convinced that we will recover as a powerful technology community. To do so, we do need to learn from the success stories of those who have prospered, but we also need to consider and try new ways and new models. In this article, I have put forward a short list of joint initiatives that haven't yet received a lot of publicity in our part of the world – let's put a few more of these options on the table, let's choose, let's build, and let's go.

References

Alberts, S. (2003). US slams the door on Canada. *Ottawa Citizen*, December 10. A9.

Bagnall, J. (2003). Lost champions. *Ottawa Citizen*, August 22. A1-A2.

Bagnall, J. (2003). Advanced attrition. *Ottawa Citizen*, August 28. F2-F3.

Bagnall, J. (2004). Cold choice. *Ottawa Citizen*, Thursday, April 22. F1.

Balu, R. (2000). All the right connections. *Fast Company*, *35*, June.

Greenaway, N. (2003). US plans to shut out Canadian defence firms. *Ottawa Citizen*, July 9. A3.

Hill, B. (2005). Nortel, Chinese firm create joint venture to make wireless gear. *Ottawa Citizen*, January 21. E1.

Moore, G.A. (1999). *Inside the tornado*. Harper Perennial.

Moore, G.A. (1999). *Crossing the chasm*. Harper Perennial.

Smilor, R., Gill, M. (1986). *The new business incubator*. Lexington, MA: Lexington Books.

Sutcliffe, M. (2005). 100,000 Tech Jobs by '08: OCRI. *Ottawa Citizen*. January 22.

Tsaprailis, E. (2004). US regions, Ottawa firms in courtship dance. *Ottawa Business Journal*, September 6. 9.

CHAPTER 7

Total Creative Revolution

Innovation and Governance at KAO Infosystems Canada

Christopher WILSON

Managing Partner, Invenire & Research Fellow, School of Political Studies, University of Ottawa

Introduction[1]

For modern companies, global competition forces an unforgiving regime of continuous change and innovation. The 'new economy' confirms the assertion by the economist Joseph Schumpeter that innovation, rather than price, drives competition. Among computer, information and telecommunications companies, this understanding is common currency, but it is less recognised in basic manufacturing operations. In addition, it is often assumed that innovation is exclusively concerned with new products, ignoring the crucial roles played by the development of new manufacturing processes, of new organisational styles and of new markets. For today's managers, therefore, fostering all forms of innovation must be an all-encompassing task to maintain competitiveness.

Typically, it is also assumed that innovation is the result of an individual genius in a lab or some young whiz kids in their garage. That perspective is complemented by the image of serendipitous discovery – the sudden flash of insight that changes everything. Increasingly, however, innovation is being understood as a social process, the product of

1 The author would particularly like to acknowledge the contributions of Craig Cunningham, Steve Nicholson and John Rajsic of KAO Infosystems Canada (now Zomax Industries). Their willingness to participate in so many open and frank discussions about KAO and to share their experience with others has been the basis of this case. I would also like to thank Gilles Paquet and Alf Chaiton for their editorial contributions and Danna Campbell for her help in preparing this monograph.

many perspectives being brought to bear on an existing problem. The project team or business alliance is *the* industry standard by which firms engage in collective learning. We can see in the Cisco vs. Nortel fight or the Wintel vs. Sun-Netscape-AOL battles that the coordination of these collective processes is an issue of major strategic importance. Understandably, the challenge of innovation becomes inevitably linked to the challenges of distributed governance, or how individuals and groups work together, including, their establishment of shared vision, effective communication, common understanding, reliable commitment and trust.

Simply put, innovation is dependent on the ability of employees and business partners to construct effective systems of collective learning. Senge (1990) and others have developed a rich understanding of organisational learning in this regard and Moore (1996), Storper (1997) and others have tried to map the nature and competencies for sustaining networked production systems. The purpose of this chapter is to juxtapose these issues of organisational learning and governance side by side, through the experience of an extremely successful manufacturing plant, KAO Infosystems of Arnprior, Canada.

During the period covered by our case writing, KAO's diskette and CD products were not subject to particularly radical change, although historically they had been. However, innovative manufacturing processes and organisational structures were required to meet the demands outlined in the case. As such, KAO's experience may have more relevance to managers in traditional industries than the experiences of computer and telecommunications related firms where product innovations tend to dominate.

KAO's Total Creative Revolution

In late April 1998 KAO's General Manager, Craig Cunningham, confirmed to his staff that the company had signed a major CD contract with one of the world's leading software manufacturers. At the same time he announced that the production equipment required to fulfill that order would be moved from KAO's Plymouth, Massachusetts plant to Arnprior. The combined news meant that the limits of both plant and personnel were going to be severely stretched. The new CD contract was to manage subscription service updates for a wide assortment of some 30-40,000 programmers and developers across North America. The task involved regular, customised, short-run production outputs and significant amounts of manual packaging.

As Engineering Production Manager, Steve Nicholson was responsible for squeezing as much efficiency and quality as possible out of

KAO's production process. "Our main problem at the time," he said, "was that we had two things overlapping – a large CD order and the equipment to fill that order needed to be moved from Plymouth." He was confident that his people were up to the new task, only he wasn't initially sure how they were going to do it. "After all," he mused, "how do you generate additional capacity in a production system that was already operating at 125% of design?" Similarly, John Rajsic, KAO's Production Manager, had to coordinate the transfer of the equipment and get the new lines up and running. In addition, Rajsic had to ensure that existing orders, accounting for 70% of KAO's current capacity, were not neglected and that a long planned capacity improvement initiative continued to move forward. Rajsic said

> The main problem from my position was the need to settle what equipment would come up. We didn't know how many lines or what condition they would be in. Eventually, we discovered that the 'new' equipment wasn't maintained all that well by our sister plant in Plymouth. So we had a lot of equipment repair and maintenance to do as well. However, once it was all installed, we ended up producing 55% more CD's than we did the previous month.

Because the new contract would involve significant amounts of manual packaging, Rajsic expected he would need a lot of casual labourers, many of whom would be entirely new to KAO. It was clear to him that some amount of training would be required. But when would that happen? So once the equipment was in place, the big question would be how to get the 'right' people – those who would maintain KAO's high standard of quality?

Another interesting twist was that the contract was being shared with KAO's Fremont, California plant. Since the West Coast operation had obtained the contract, they would be the interface with the client. The challenge, therefore, for KAO Canada would be to present to the client a seamless output in terms of product and service. That meant there had to be a high degree of synchronisation with Fremont.

KAO Infosystems Canada

KAO Infosystems Canada (KIC) is the Canadian manufacturer for the Infosystems division of KAO Corporation of Japan. In 1997, the KAO Corporation generated almost $8 billion (US) in revenues, of which the Infosystems Division contributed sales of over $900 million. KAO's original core competency in surficants had made it a world leader in consumer products with such household brands as Jergen's soap, Biore personal care products and Goldwell cosmetics. But that same expertise also gave KAO a natural advantage in the production of data storage media where surficant technology is a critical success factor. In 1998, KAO Infosystems manufactured magnetic and optical media such as computer diskettes (floppy disks), CD-ROM's, CD recordables, and DVD's.

In addition, KAO Infosystems provided customised production and world-wide distribution services for content developers that included CD replication, diskette duplication, end-to-end turnkey services, as well as marketing, channel development and customer services. KAO Infosystems operated eight production facilities worldwide, including the Arnprior, Ontario plant, and all were ISO-9002 certified. In 1996 KAO Infosystems was ranked as the world's number one CD-ROM producer, and in 1997 produced 175 million CD's.

KAO Infosystems Canada began in 1983 as Didak Manufacturing and was located in Arnprior just outside of Canada's national capital: Ottawa. At the time, Didak produced 100,000 5¼-inch computer diskettes per month under the 'Axiom' brand name. A merger between Didak and the Japanese firm KAO in 1986 resulted in KAO-Didak and the quadrupling of its manufacturing space. The merger added state-of-the-art production equipment that automated the production process from start to finish. From this point, KAO-Didak operated 24 hours per day, seven days a week and was producing 2.5 million 3½-inch diskettes per month. The company was again reorganised in 1993 as KAO Infosystems Canada (KIC), and the Arnprior plant went from an 80,000 to 140,000-sq. ft. production facility resulting in a tripling of diskette output per month. In 1994 KIC became ISO certified – a challenge in itself for any business, but more so in light of their move into an entirely different media business – CD manufacturing.

With the addition of another production line in 1996, KAO increased diskette production output to its 1998 level of 10 million pieces per month, but it also doubled CD production and added the capacity for CD glass-mastering. While floppy disks could be mass-produced and inventoried, the CD business was extremely customised, with some runs as low as several thousand. Thus, CD's added another altogether different level of complexity to KIC's operation.

In keeping with its tradition of providing end-to-end services, KAO also added MPEG and video production studio services in Toronto. In April of 1998, in an effort to streamline their operations, KAO Infosystems closed their Plymouth, Massachusetts plant and relocated diskette-marketing functions for the eastern US market to Arnprior. In addition, two additional CD production lines were to be moved north – a move that was to be completed by mid-May. The resulting increase in CD capacity from 1.8 to 2.7 million pieces per month was a major factor in the new contract, but Nicholson's challenge was the small window in which to move the equipment and train the staff.

In the twelve years since Didak had merged with KAO, the data storage industry had undergone dramatic changes – changes in product, changes in production, changes in markets, and changes in response time – all on a background of increasingly cut-throat competition and declining margins. Instead of producing the simple but technologically sophisticated product of the mid-1980s, by 1998 KAO was manufacturing a variety of products, both magnetic and optical. It offered a variety of services from product registration and customer information services to the bundling of OEM software products to global marketing solutions that included direct marketing, Internet commerce, distribution and upgrade management.

KAO had also moved from a supply-based system of production to a production-on-demand style that had reduced their response time from months to weeks or often to hours. Customer profiles had changed from the original independent software vendors to become OEM hardware vendors, publishers of a wide variety of previously printed materials, and suppliers of both entertainment and edutainment multimedia products. Both diskettes and CD-ROM's had become commodity items, their distinguishing features becoming principally price and quality, although the best quality was now largely assumed. Consequently, the industry had witnessed a veritable free fall in price. Diskette prices, for example, had dropped an average of 23% per year between 1986 and 1996. Where diskettes once sold for 2 Cdn$ – 2.50 Cdn$ each, they now wholesaled for less than 20¢. Consequently, cost control and process innovation to reduce the cost of production became central to the organisation's survival.

KAO Infosystems Canada responded to these challenges in three ways:

- First, they instituted programmes of continuous improvement and innovation;
- Second, they developed a skilled, stable workforce either by recruitment or training; and,
- Third, they created a learning culture that ensured the requisite adaptability for the company in the face of dynamic industry change.

KIC's approach resulted in a gradual production increase to a level that was 125% of their equipment's design capacity and a doubling of per capita output between 1993 and 1998. KIC's employee turnover rate was 3-5 times lower than that of other high-tech firms in the Ottawa area, even though their wage rates were lower than industry leaders. Their measure of innovation was Total Cost Reduction or TCR, which KAO re-defined as Total Creative Revolution. KIC achieved TCR rates of 50% in one year. In 1997, KIC was ranked as the leading plant within the industry's leading company (in terms of efficiency and quality). Just as impressive was the company's client list that included such leaders as Microsoft, IBM, HP, Apple and Lotus.

Innovation & Learning at KAO

In order to respond to the relentless competitive demands for lower prices, KAO's management was quick to adopt a philosophy of embracing continuous learning and improvement. As KAO's understanding of market conditions evolved, that philosophy began to shape the coordination mechanisms necessary to meet the triple challenges of innovation, retaining a skilled workforce, and corporate flexibility.

According to KIC's General Manager Craig Cunningham,

> We learned that philosophy and empowerment were not enough. We also learned that developing skills and experience were not enough. As we saw it, the key to continuous improvement was knowledge. The more knowledgeable our employees became, the more perspectives they could bring to bear on our production process. So we set about developing a learning culture at KAO and building our own knowledge bank.

To create a learning culture at KAO, senior managers promoted three major principles throughout the organisation. They were:

1. everyone must learn
2. everyone must teach
3. advancement was linked to knowledge.

The first principle, *everyone must learn*, applied to everyone from line workers to senior executives. Learning was viewed as an integral part of everyone's job description. Not only were employees expected to perform specific tasks but also a certain amount of job time was to be dedicated to learning. The second principle, *everyone must teach*, enshrined the idea that knowledge was to be shared, not hoarded, among others in the organisation. Since teaching was based on knowing and not on formal positions this second principle did much to remove barriers between people and move responsibility downward in the organisation. The third principle, *advancement was linked to knowledge*, tried to create the proper incentives throughout the organisation. For example, job expectations were tied to knowledge acquisition. If employees wanted to develop a career path within KAO, that path would be linked to their willingness to learn and broaden their knowledge base. If they wanted a merit increase, that increase was tied to their ability to apply new knowledge or demonstrate their innovativeness. If they wanted secure employment, that security was connected to their increased capability to contribute to KAO.

To facilitate learning, KIC initially undertook to create a series of formalised courses, offering everything from machine operation, to statistical process control, to computer applications, to the understanding of financial statements. Drawing faculty mainly from KIC employees, KIC established *KAO Infosystems Canada University* (KICU). Because employee learning became generally recognised as a critical success factor, KAO employees were given time off *with pay* to go to school.

At its height, KICU offered over 80 different courses. "We went way overboard in the number of courses we offered," said Cunningham, "but we needed to test the limits of where this idea would lead us." By 1998 KICU had been scaled back a bit but still about two dozen courses were offered, including advanced topics in management – sales & marketing, project management, and inventory management – as well as technical courses in computer programming, electronics, pneumatics, and computer networking.

The precedents set by KICU also had an impact on external training programmes. Courses not available internally through KICU were made accessible to employees. The philosophy was that employees should be learning even if that learning wasn't obviously linked to their job description. Employees were encouraged to seek diploma courses to advance their knowledge and that training would be subsidised with tuition grants and time off. However, there was a price employees had to pay for learning externally – they were expected to share their newfound knowledge on their return, often by setting up another internal KICU course on the topic.

While employee learning was generally self-directed, some courses, such as those on company structure and philosophy, statistical process control (SPC), and ISO 9002, were required for all company employees. This presented some unique challenges, particularly with SPC. SPC was required because the company wanted to show that every corporate process could be broken down, measured, analyzed and improved upon. However, SPC demanded a minimum level of mathematical understanding and for many KIC employees who had only rudimentary high school math this requirement appeared daunting. KIC management though, responded with patience. In one case, it took an employee four attempts to pass the SPC exam. Such patience, however, paid off by demonstrating the firm's commitment both to the need for SPC and to its employee's growth.

This tolerance of failure became a key element of KIC's culture. Cunningham said

> If you want to innovate then you have to expect that sometimes people won't get things right the first time. This is the downside of a learning culture that is often ignored. A lot of times I, or one of my managers, have been awoken in the middle of the night because a machine operator tried to apply what he'd learned in the classroom and broke the machine. Sure it's inconvenient as hell to come in at four in the morning to fix a machine, but in the long run, that willingness to apply something new has been essential to our competitiveness. Without it we would have been surpassed long ago.

KAO's principal means of fostering innovation has been in its creation of an environment in which both learning and creativity could flourish. KAO managers encouraged employees at every level to take the initiative and improve upon both products and processes. This was done in part by selecting people who were very 'hands-on' types, who saw problems as challenges and were willing to tackle them. Steve Nicholson said

> I'm sure it is surprising to some that KAO Infosystems, this very high-tech, very automated manufacturer is located way out in rural Arnprior. I mean we have cornfields just across the road. But it turns out that farmers tend to be our best employees. They aren't afraid of getting their hands dirty. They have this inbred attitude that if there is a problem you fix it. You don't wait – you just do it. I hired someone not too long ago who showed up late to his interview dressed in jeans and looking a little rough. I know that elsewhere he would not have been given a second thought. But he'd had a lot of experience fixing things on his own. I had this gut feeling that he would work out and he did. I often find him trying to improve upon things on his own.

Once on the job, employees discovered that it's OK to modify things or work outside their specific area of responsibility. In fact, they soon came to recognise interesting ideas can readily give them access to all sorts of 'experts' from across the company who were willing to help them. At KIC, 'skunkwork projects,' the kind of informal R&D projects that go on outside the general direction of managers, were common. "You know I'm an engineer by profession," said Cunningham, "and there's nothing I like more than running across a group of people trying to figure out a new way of doing things. I can't tell you how many times I've gotten sucked into one of these groups. It's fun for me and a good way for me to keep in touch with what's going on."

"If you go to a firm like Nortel, you find a lot of people who work as specialists" said Nicholson. "Here people are encouraged to acquire a very broad knowledge about everything we do. We do a lot of cross training. People bounce around a lot from one job to another. It's never dull and people seem to derive a certain satisfaction from never knowing what it is they'll be doing that day. It makes work fun." Recalling that KAO operated in two different businesses, floppy disks and CD's, Nicholson said, "the creation of a liquid workforce that could operate between the two is an essential management goal."

One thing KAO Infosystems didn't skimp on was praise. Besides a regular pat on the back, KAO created what employees refer to as the 'Board of Fame' to publicly recognise innovative employees. "People respond well to positive feedback, more so even than to money," said Cunningham. "We put up the 'Board of Fame' right next to where we usually meet with new clients in order to give them a sense of our ongoing creativity. As it turned out, this became a major selling tool for us." KAO also maintained a box at the Corel Centre for sports and special entertainment events but instead of using it as an executive perk, it was used almost exclusively to reward employees who had made innovative contributions to the firm.

Unlike other organisations which tended to give lip service to creating both learning and innovation cultures, KAO was pretty free with learning's essential currency – information. Employees could generally access whatever information they wanted on the company's computer network. The company maintained an open library and encouraged employees to seek out and maintain outside contacts both within and outside of KAO's far-flung global organisation.

Since the organisation understood that learning could take place anywhere, managers, mid-level engineers and designers made sure that they took the time to explain to line operators and technicians why they did what they did. Being accessible was another trait of KAO managers. "It is a trait we hire for," said Cunningham. "If someone has all kinds of degrees and management experience but is not willing to get down in the trenches to teach and learn then that person isn't for us." Problems at KAO were not viewed as occasions to lay blame but as opportunities to correct and learn.

As an example of KAO's ability to conduct joint problem solving, it periodically ran a 'Capacity Improvement initiative' that involved small teams who had an interest in improving upon certain production performance issues. The teams developed and implemented their own changes, then monitored and fine-tuned them. The teams were drawn from across the company – operators, technicians, engineers and support staff. Team leaders could be anyone.

In fact, Cunningham cited the experience of one employee who had been regarded as "on his way out" but who took responsibility for an improvement team because he felt particularly bugged by the problem. When this person learned that as team leader he could get access to whatever information and expertise he needed to solve the problem, his innate curiosity seemed to turn on and he started talking to technicians, engineers and even finance people. When that didn't satisfy him, he started taking KICU courses and when that wasn't enough he took outside courses. "In the end," said Cunningham, "his team came up with a very effective solution, but more important than that we gained someone who is now one of our top employees."

From the Capacity Improvement initiative, engineers learned from line personnel and vice versa. Morale improved because of the real sense of empowerment and communication across the organisation. "The CI initiative really got people working together and it gave them ownership of what we do here," said Nicholson. "Keeping people interested is a key challenge for me. CI focussed people on specific tasks and then all we had to do was keep out of their way. The result was that they are now more conscious and vigilant in their work."

Governance & Coordination at KAO

As John Rajsic would point out, although the Production Manager's job did involve keeping equipment running, his main task was to create an adaptive capacity among KAO's employees to meet any challenge that came along. "To get the best out of people, you really need to get to know them. It's not just what they say, but it's in their body language and tone of voice as well." Rajsic's appreciation of this fact came from a twenty-year career in manufacturing which he began as an industrial mechanic and which later found him in management after honing his people skills in listening, communicating and mentoring.

In 1993, when Rajsic was hired as Production Supervisor at KAO, he was attracted by the opportunities for learning and professional growth – a fact reflected in the thirty-odd courses he has taken since then. In October 1996, Rajsic became Production Manager with particular responsibility for KAO's CD operations.

According to Rajsic, coordination at KAO was very future oriented and was based on three elements – healthy relationships, frequent communications, and the building of an adaptive capacity to respond. "Reducing our changeover times is crucial to our business. We operate with a 'pit-stop mentality' here. We have to know what to do and to be able to do it quickly or the time we lose could cost us the race." Consequently, Rajsic often found himself trying to prepare for challenges that had not yet arisen rather than dealing with the ones that had been faced before.

The first component of KAO's strategy, 'healthy relationships,' reflected the company's clear recognition that although people could be a huge resource, they were turned into assets only by continually nourishing the company's relationship with them over time. "I'm continually talking to people," says Rajsic. "I need to know what my staff can do and what their limits are. But they need to know that too, so we talk about what works on the job and what doesn't. When we find out what's missing, then I work with them to develop their skills. With most people this isn't an open-ended thing. The Peter Principle still applies. But once we can agree what the limits are, I let them do their job. Successful management," Rajsic emphasised, "doesn't mean constant monitoring. Only a bad manager has to be there all the time."

Once again the key element in building those relationships within KAO lay in the organisation's encouragement of learning and the sharing of knowledge. The cross cutting of the teaching and learning experience among employees at all levels contributed immensely to the reduction of knowledge hierarchies and the 'stove pipe' style that many organisations seemed to operate with.

As it turned out, healthy relationships were important not only within KAO but also between the Arnprior plant and its external stakeholders. KIC maintained very close connections with its sister plants within KAO. It was frequently the case that the Arnprior plant would be asked to produce all or part of an order generated from any one of KAO's principal locations in the US, Ireland, Japan, or Spain, or for that matter from any of its 55 offices worldwide. The establishment of working relationships with their counterparts in these other organisations became an essential element of KIC's problem solving capacity. As Rajsic put it, "when there is a problem, people are much more forgiving if a relationship is already established." To create those relationships, Rajsic relied on regular face-to-face sessions to build the needed trust and understanding.

The second component of KAO's coordination strategy was frequent communications. While information tended to be available to whoever needed it, the more complex projects tended to receive more frequent communication to head off potential problems. Each morning KAO executives would meet to prioritise jobs for the day. As production problems or new considerations arose during the day, orders could be delayed or moved forward in priority, based on a three-dimensional rating system that included loyalty, revenue and strategic importance.

The scheduling process was a dynamic one that could be changed as soon as it took to make a decision and do the changeover – usually within an hour.

A scaled pricing system with dimensions of time urgency and size was also put in place. For example, those customers who wanted to drop off their CD content at closing time on one day and pick it up next morning and had fewer than 1,000 copies paid the highest price. The least costly jobs allowed 10 days or more turnaround on tens of thousands of copies. The pricing system thus helped KAO to prioritise its schedule as well as to earn a profit.

The third component of KAO's coordination strategy was aimed at building a capacity to respond to the high levels of uncertainty they were often faced with. When scheduling decisions were made, production capacity was always buffered and downtime budgeted. "We always try to ensure at least 15% excess production capacity to handle the inevitable breakdown of equipment or the urgent last minute order," said Rajsic. "We have a hard time turning away the small, loyal customers who have supported us all these years. Their business has often meant the difference between being profitable or not."

KAO also took advantage of the fact that they operated in two different manufacturing businesses – diskettes and CD's. The diskette business was a high-volume, automated process and so it was natural to build up inventories to achieve the necessary scale economies. The CD business, on the other hand, was more of a short-run, customised operation that was consequently more labour-intensive. When diskette inventories were high, employees from the diskette side were cross-trained in CD production during slack periods. Almost 50% of KIC's staff were cross-trained in this manner. The result was a very fluid workforce that could move as demand dictated. "It's like having 8-10 extra people just floating and able to fill in the gaps," said Rajsic. KIC would also use slack periods to cross-train its people with its sister plants in the USA and Europe. Because of its reputation for innovation and quality, KIC also became a preferred site for KAO employees from other sites to come and train.

Building adaptive capacity also involved creating a virtual labour force, a casual labour pool that KAO had been developing over time that had experience with KAO's operations. The size of this group ranged from 65 - 165 people and had a full-time coordinator within the Human Resources Department. While many customers shut their eyes to it, it was a common practice in the industry to outsource jobs to competitors that for one reason or another could not be entirely handled by the main producer. In KAO's case, several such competitors had become occasional partners learning to deliver the same quality that had made KAO an industry leader. These relationships were forged over many years of trial and error in the accumulation of trust capital.

Options & Solutions

As both Nicholson and Rajsic began dealing with the governance challenge of innovation, under the dual constraints of intense time pressure and complex interrelated tasks, they recognised a number of options. They could hire additional people to run the new CD lines, but many of the tasks that would be required of them would be very specialised. Since no company in the Ottawa Valley came up to KAO's level of automation, this option entailed having to spend a lot of time training. Time, however, was something that they did not have. Alternatively, they could allow more overtime, but that would have had a big impact on costs and impair the effectiveness of their people. They could also contract the new CD job out to another firm but it would be hard to find anyone who could do a job of this size on such short notice. In addition, the client was clearly expecting KAO's highest level of quality. Further, with the larger corporate responsibilities being established with the Arnprior plant, it would be unseemly to put this job off onto someone else at this time.

KAO's 'solution' was to take advantage of the flexibility they had built up in their people. First, a 'crack' team from the existing CD line was sent to Fremont, California to better understand the needs of their client. Second, another team was sent to Plymouth to oversee the operation and dismantling of the equipment down there. Since KIC had extensively cross-trained about half of their diskette staff in the CD operation, diskette inventory was quickly ramped up and then the 'excess' personnel were shifted over to CD production.

KAO's major concern centred on the large staffing needs at the packaging end of the process. The casual labour pool of experienced workers proved to be invaluable. While additional staffing requirements were outsourced to Olsten, a temporary help agency, KAO's more experienced casual workforce supplemented their lack of training.

To ensure high customer satisfaction, the synchronisation between Fremont and Arnprior in terms of quality and timing of outputs began early. In preparation for the new contract rollout, Rajsic travelled many times to California to make sure he and his colleagues in Fremont were all on *the same page*. In the time before the production lines were set up, the Arnprior management team began meeting weekly to discuss the progress of the various teams and the Fremont group was also included via teleconferencing. This minimised the chance of any surprises to the whole network of players. As production started, these meetings increased in frequency and became daily progress updates.

In the end, even though the 'new' equipment was in need of greater repair than anticipated and the client was itself almost two weeks late with its software masters, KAO was able to absorb these additional 'shocks' and still deliver to the client's customers on time. It was this type of adaptability that had made KAO's Arnprior plant number one in the industry.

Conclusions

The dilemmas faced by both principals in the case, Steve Nicholson and John Rajsic, illustrate a common challenge faced by many modern organisations responding to global competitiveness – how do you constantly innovate, under the dual constraints of intense time pressure and complex interrelated tasks? In KAO's case, the company had to deal with the challenge of equipment upgrades overlapping a large new contract for which production had to begin within a narrow one-month window. The company's response was embedded in their culture of continuous learning and flexible governance that enabled it to shift and add resources, as the need required. "The bottom line," according to Rajsic "was that we had only one month to do it all – to get the equipment, repair it, install it, find the right people, hire and train them to process and package the order, and in the end provide our client with a level of quality and service which was beyond their expectations."

A very clear message from the experience of KAO Infosystems Canada was the importance of collective learning. That learning took many forms but its ubiquitous presence provided KAO with a capacity for change that was and continues to be uncommon in most organisations. Clear, too, was the centrality of the human relationships among managers, employees and partners and the mutual respect each accorded the other. It was evident that these relationships had been institutionalised in a variety of ways so as to optimise the degree of innovation and adaptability within the organisation and provide a self-governing framework similar to one described by Hayek (1989) that motivated "individuals to do the desirable thing without anyone having to tell them what to do." This is the essence of economy – doing less and accomplishing more. For organisations that want to succeed in the new global marketplace, these are critically important lessons.

Epilogue

In spite of KAO's success in production innovation, the overall decline in diskette prices continued to minimise the contribution from diskette manufacturing to KAO Infosystems' profitability. In September 1998, KAO Corporation of Japan decided to concentrate on its core competencies in household products and eliminate its Infosystems division. Subsequently, KAO Infosystems Canada sold off their diskette production facilities to make it more attractive to potential buyers. The move shed some 120 local jobs. It is significant that within weeks other area technology firms had hired all of those laid off. KAO offered to provide part-time employment to ease any transition pain its employees might experience but their offer was never taken advantage of.

On December 2, 1998 KAO announced the sale of its Infosystems division to Zomax Industries of Minneapolis for 60 million Cdn$. The acquired Kao Infosystems assets and facilities included a call centre in San Ramon, CA, a twenty-two line CD and DVD manufacturing, packaging and distribution facility in Fremont, CA and the six line CD manufacturing, packaging and distribution facility in Arnprior, Canada. The purchase also included KAO's Irish assets – a call centre in Dublin, an eleven line CD manufacturing, packaging and distribution facility in Dublin and a diskette duplicating, packaging and distribution facility in Langen, Germany.

References

Hayek, F. von. (1989). The pretence of knowledge, Nobel memorial lecture, December 11, 1974. *American Economic Review*, *79* (6), 3-5.

Moore, J.F. (1996). *The death of competition - leadership and strategy in the age of business ecosystems*. New York: HarperCollins.

Senge, P. (1990). *The fifth discipline*. New York: Doubleday.

Storper, M. (1997). *The regional world: territorial development in a global economy*. New York: Guilford Press.

CHAPTER 8

Comparing Science and Engineering Employment in Canadian and US Cities

Desmond BECKSTEAD and W. Mark BROWN

Micro-Economic Analysis Division,
Statistics Canada

Introduction[1]

It has long been argued that one of the keys to economic growth is investment in human capital (Lucas, 1988). In Canada, there is evidence that increases in human capital are associated with higher levels of productivity growth, which in turn is an important contributor to long-run economic growth (Gu *et al.*, 2003). Human capital can be measured in several ways. Educational attainment combined with years of working experience is the most commonly used method of quantifying human capital.

In this chapter, we take a simpler approach and identify individuals who work in occupations that require large investments in human capital. Scientists and engineers figure prominently in this regard. Moreover, their activities are seen by many as essential to technological innovation and economic growth. For instance, the Progressive Policy Institute begins its analysis of changes in engineering and science employment by noting that "[t]echnological innovation is one of the key drivers of overall economic progress, and it is fuelled by a strong engineering and scientific workforce" (Atkinson and Court, 1998, p. 41). Similarly, the National Science Foundation (2004, Chapter 3, p. 5) notes that scientists and engineers "contribute enormously to technological innovation and economic growth, research, and increased knowledge." Given the importance accorded to scientists and engineers as drivers of

1 The authors are grateful to John R. Baldwin and Guy Gellatly, among others.

technological change and economic growth, we limit our measure to workers in science and engineering (S&E) occupations.

In recent years, cities have become increasingly interested in attracting human capital.[2] At least in part, this interest stems from the positive effect that high levels of human capital can have on future growth (Glaeser and Saiz, 2004). Naturally following from this interest is a need to find a basis of comparison; that is, to see how individual cities are performing relative to each other. The purpose of this chapter is to provide such a comparison.

As Canada is increasingly integrated into a North American labour market for skilled professionals,[3] Canadian cities have begun to compete on a continent-wide basis. Hence, it is necessary to compare the performance of Canadian cities to US cities.[4] To this end, this chapter provides an evaluation of the relative strength of employment in science- and engineering-based occupations in cities across Canada and the United States.

Performance is measured in two ways. First, Canadian and US cities are compared based on the share of their employed population in S&E occupations in the early 2000s. Here we use the most recently available data to determine whether S&E employment is more or less important in Canadian and US cities. Many analysts are interested not only in how Canadian cities fare in general, but also how individual metropolitan areas perform. To this end, we present the S&E employment shares of individual Canadian cities and how they rank on a North American basis.

The mechanics of how we go about comparing cities, be they in Canada or in the United States, is important. There is a growing literature[5] that suggests larger cities tend to have higher shares of workers with highly specialised skills, like scientists and engineers. Hence, we not only show how Canadian cities rank compared to all cities in North America, but also how they compare to their peer cities of similar size (e.g., Toronto and Boston, or Halifax and Ann Arbor).

---

2 See Florida (2002a, 2002b) and Gertler *et al.* (2002).

3 Devoretz and Coulombe (2005, p. 436) in their extensive review of the evidence on North American integration conclude that there is substantial labour market integration for selected occupations.

4 Recent comparisons of the economic and social characteristics of Canadian and US metropolitan areas can be found in Gertler *et al.* (2002) and Polèse and Tremblay (2005).

5 This literature is discussed in more detail in Section 4.2.

Second, in addition to providing a snapshot of Canadian and US cities at the turn of the last century, we also compared them in terms of their performance over the previous 20 years. As a result, we are able to determine whether the trend in S&E employment shares is favourable or unfavourable for Canadian cities. We also ask which cities have experienced the strongest increase in their S&E employment shares over the past two decades.

The remainder of the chapter is organised as follows. In the next section (Section 2), we describe the sources of the data for the analysis and how we define science and engineering occupations. This is followed by a brief overview of how the Canadian and US labour forces compare in terms of S&E employment in the 1980 and 1981 period and the 2000 and 2001 period (Section 3). After this, the chapter focuses on how Canadian and US cities compare. The analysis is divided into two sections that address first how Canadian and US cities compare in 2000 and 2001 (Section 4) and second, how they have changed over time between the 1980 and 1981 period and the 2000 and 2001 period (Section 5). The chapter ends with a brief conclusion (Section 6).

Data

The data for this project are derived from the 1980 and 2000 US censuses and the 1981 and 2001 Canadian censuses.[6] Both censuses provide information on the location and occupation of workers. The locational information allows us to link workers to specific cities, while the occupational information allows us to identify scientists and engineers.[7]

We base our definition of scientists and engineers on the National Science Foundation's S&E occupational framework. The S&E occupations are drawn from five categories: computer and mathematical scientists, life and related scientists, physical and related scientists, social and related scientists, and engineers. Table 1 outlines a more detailed description of these occupations.

---

6 Since we require occupational information, we need data from the more detailed 'long forms' that are distributed to approximately every one in six households in the United States, and every one in five in Canada (Census 2B). For the US data, we used the PUMS data file, which is composed of a sample drawn from the one in six household sample that amounts to 5% of the US population.

7 See Beckstead and Gellatly (2006) for a discussion of alternate measures of the incidence of scientists and engineers.

Table 1. Inventory of Science and Engineering Occupations

Computer and mathematical scientists	
	Computer and information scientists
	Mathematical scientists
Life and related scientists	
	Agricultural and food scientists
	Biological and medical scientists
	Environmental life scientists
Physical and related scientists	
	Chemists, except biochemists
	Earth scientists, geologists and oceanographers
	Physicists and astronomers
	Other physical and related scientists
Social and related scientists	
	Economists
	Political scientists
	Psychologists
	Sociologists and anthropologists
	Other social and related scientists
Engineers	
	Aerospace and related engineers
	Chemical engineers
	Civil and architectural engineers
	Electrical and related engineers
	Industrial engineers
	Mechanical engineers
	Other engineers

Canadian cities are defined as Census Metropolitan Areas (CMA) and Census Agglomerations (CA) as defined by Statistics Canada. US cities are defined using the Metropolitan Statistical Area (MSA) definition of cities utilised by the US Census Bureau. Since MSAs are limited to those areas with a population of 100,000 or more we exclude Canadian CMAs/CAs with a population less than 100,000. In 2000 and 2001, of the 316 metropolitan areas in our sample, 282 are US and 34 are Canadian. For ease of exposition, throughout the chapter we describe CMAs/CAs and MSAs as simply 'cities.'

Change in Science and Engineering and Related Employment between 1981 and 2001

The proportion of employed workers in S&E occupations in Canada and the United States both increased between the 1980 and 1981 period and the 2000 and 2001 period (Table 2). In 1980, 2.6% of employed workers in the United States worked in S&E occupations, while in 1981 this share was 2.3% for Canada. By 2000, the share of workers in the United States in S&E occupations had increased to 4.5% and this share was matched by Canada.[8]

Table 2. Canada and United States Science and Engineering Employment Shares, for the Years 1980, 1981 and 2000, 2001 (percentage)

	1980 and 1981	2000 and 2001
Canada	2.3	4.5
United States	2.6	4.5

Source: Canadian censuses (1981 and 2001) and US censuses (1980 and 2000).

While Table 2 provides us with an overall picture of the importance of S&E occupations, it does not provide a picture of how cities in general have performed. We do not know whether S&E occupations are more prevalent in larger urban centres relative to smaller centres and rural areas nor whether employment has become more concentrated in larger urban centres over time. To answer these basic questions, we divide employed workers into two groups: (1) workers who are living in relatively large cities with a population of 100,000 and more (Cities) and (2) workers who are living in smaller cities and rural areas (Small Urban-Rural). Table 3 presents the results of this breakdown.

Science and engineering employment accounts for a larger share of employment in Cities than in Small Urban-Rural areas in both countries and this gap has widened over time. In 1980, the share of employment in cities was roughly twice that of Small Urban-Rural areas in the United States at 3.0% and 1.6%, respectively. Cities in the United States increased their S&E share of employment by 2.1 percentage points between 1980 and 2000, while Small Urban-Rural areas increased their share by only 0.7 percentage points. The same pattern was repeated for Canada. There appears to be an urban focus in terms of the incidence

8 See Beckstead and Gellatly (2006) for a more detailed discussion of trends in S&E employment at the national scale.

and the growth in science and engineering occupations. This is a theme that we return to in the two subsequent sections.

Table 3 also provides evidence that much of the difference in S&E employment shares between Canada and the United States in 1980 and 1981 was due to relatively low levels of S&E employment in Small Urban-Rural areas in Canada. In Small Urban-Rural areas in the United States, the incidence of S&E employment was 1.6%. It was 1.3% in Canada. This gap continued to be present in 2000 and 2001.

Table 3. Cities and Small Urban-Rural Science and Engineering Shares of Employment, for the Years 1980, 1981 and 2000, 2001 (percentage)

	1980 and 1981		2000 and 2001	
	Cities	Small urban-rural	Cities	Small urban-rural
Canada	2.9	1.3	5.6	2.0
United States	3.0	1.6	5.1	2.3

Source: Canadian censuses (1981 and 2001) and US censuses (1980 and 2000).

In contrast to Small Urban-Rural areas, Canadian cities have performed well relative to their US counterparts. In 1980 and 1981, their share of employment in S&E occupations was just behind that of US cities and by 2000 and 2001 the S&E employment share in Canadian cities was above that of US cities.

The remainder of the chapter focuses on those metropolitan areas in Canada and the United States that fall within the Cities classification. We ask how specific Canadian cities compare to their US counterparts in 2000 and 2001 and how they have evolved since 1980 and 1981.

Comparing Canadian and US Cities in 2000 and 2001

The objective of this section is to develop a comparison of S&E employment across Canadian and US cities. The comparison is based on the absolute size of S&E employment and on the S&E share of employment in each city. The former provides a measure of which cities are the most important locations for S&E employment, while the latter helps to identify cities whose employment is concentrated in S&E occupations. The S&E share of employment also serves as a measure of the ability of urban economies to support S&E employment.

The main purpose of any city comparison is to see how cities rank relative to each other. However, as will become clear, this is not purely a matter of listing the highest ranked cities – although this is done. Rather, since there is such a strong positive relationship between city size and S&E employment shares, it is also necessary to compare cities of similar size. Arguably, this peer-based comparison provides a more appropriate measure of performance. Finally, it is important to note that, given the large number of cities in our sample, it is impossible to present them all in the chapter.[9]

Ranking Canadian and US Cities by Science and Engineering Employment

The first step in the analysis is to compare levels of S&E employment across Canadian and US cities. This provides us with a measure of how Canadian cities rank in terms of the absolute size of S&E employment. Presented in Table 4 are the levels of S&E employment for the top 30 cities of the 316 Canadian and US cities in our sample, or about the top 10%. Also reported in the table are the population and the population ranks of these same cities.

In term of absolute size, the New York, Washington and Los Angeles metropolitan areas have the largest S&E workforces. There are four Canadian cities that make the top 30. Toronto ranks seventh in terms of S&E employment, as it also does in terms of population. Montréal ranks 16th for S&E employment, slightly behind its 13th population rank. Remarkably, Ottawa-Gatineau ranks 21st in term of S&E employment, while ranking only 51st by population. Finally, Vancouver ranks 26th in term of S&E employment and population.

Absolute size provides one perspective on the variation in S&E employment across cities. Alternatively, we can observe the share of employed workers who are employed in S&E occupations – a measure of how important S&E employment is for each city. The average S&E share of employment for Canadian cities in 2001 was 4.2%, while the average for the United States is slightly lower at 4.0%. These averages are less than those presented in Table 3 (above) because they are not weighted by the level of S&E employment in each city. Smaller cities are more numerous than large cities and, as we illustrate, they tend to have lower S&E employment shares. As a result, the unweighted or arithmetic averages presented here are lower than those presented in Table 3.

9 For the interested reader, all comparative measures developed in this section are presented in Beckstead and Brown (2006) for all 316 Canadian and US cities.

Table 4. Science and Engineering (S&E) Employment Levels for the top 30 North American Cities, 2000 and 2001

	S&E Employment	Rank	Pop'n	Rank
New York-Northeastern, NJ	369,193	1	17,244,066	1
Washington, DC/MD/VA	271,804	2	4,733,359	6
Los Angeles-Long Beach, CA	239,237	3	12,368,516	2
Chicago-Gary-Lake, IL	211,198	4	8,804,453	3
San Francisco-Oakland-Vallejo, CA	184,596	5	4,645,830	8
Boston, MA	167,285	6	3,951,557	12
Toronto	*161,552*	*7*	*4,682,898*	*7*
Dallas-Fort Worth, TX	156,797	8	5,043,876	5
San José, CA	133,121	9	1,688,089	28
Philadelphia, PA/NJ	126,213	10	5,082,137	4
Detroit, MI	124,992	11	4,430,477	9
Atlanta, GA	118,578	12	3,987,990	11
Houston-Brazoria, TX	114,318	13	4,413,414	10
Seattle-Everett, WA	105,929	14	2,332,682	21
Minneapolis-St. Paul, MN	100,067	15	2,856,295	16
Montréal	*91,523*	*16*	*3,426,350*	*13*
Denver-Boulder-Longmont, CO	90,420	17	2,198,801	25
Baltimore, MD	84,918	18	2,513,661	19
San Diego, CA	78,935	19	2,807,873	17
Phoenix, AZ	75,590	20	3,070,331	15
Ottawa-Gatineau	*65,332*	*21*	*1,063,664*	*51*
Raleigh-Durham, NC	62,994	22	1,182,869	45
St. Louis, MT/IL	62,202	23	2,602,448	18
Austin, TX	56,210	24	1,167,216	47
Portland-Vancouver, OR/WA	51,372	25	1,789,019	27
Vancouver	*51,280*	*26*	*1,986,965*	*26*
Sacramento, CA	47,303	27	1,632,863	31
Pittsburgh-Beaver Valley, PA	47,224	28	2,285,064	22
Cleveland, OH	44,366	29	2,255,480	23
Columbus, OH	44,319	30	1,443,293	39

Source: Canadian Census (2001) and US Census (2000).

Table 5. Science and Engineering Employment Shares for the Top 30 North American Cities, 2000 and 2001

	Share (%)	Rank
San José, CA	15.7	1
Ottawa-Gatineau	*11.6*	*2*
Huntsville, AL	11.1	3
Nashua, NH	11.1	4
Washington, DC/MD/VA	10.9	5
Raleigh-Durham, NC	10	6
Rochester, MN	9.6	7
Ann Arbor, MI	9.2	8
Austin, TX	9	9
Santa Fe, NM	8.9	10
Seattle-Everett, WA	8.6	11
Boston, MA	8.3	12
Yolo, CA	8	13
Fort Collins-Loveland, CO	8	14
San Francisco-Oakland-Vallejo, CA	8	15
Trenton, NJ	8	16
Dutchess County, NY	7.9	17
Santa Cruz, Calif.	7.8	18
Melbourne-Titusville-Cocoa-Palm Bay, FL	7.8	19
Denver-Boulder-Longmont, CO	7.8	20
Colorado Springs, CO	7.8	21
Calgary	*7.6*	*22*
Madison, WI	7.5	23
Richland-Kennewick-Pasco, WA	7.4	24
State College, PA	7.1	25
Bloomington-Normal, IL	7	26
Baltimore, MD	6.9	27
Wilmington, DE/NJ/MD	6.9	28
Champaign-Urbana-Rantoul, Ill.	6.7	29
Toronto	*6.7*	*30*

Source: Canadian Census (2001) and US Census (2000).

The top 30 cities ranked by share are presented in Table 5. Three Canadian cities make the top 30 in terms of S&E employment shares – Ottawa-Gatineau, Calgary and Toronto. Ottawa-Gatineau ranks second only to San José, which encompasses much of 'Silicon Valley.'[10] Ottawa-Gatineau's high placing suggests its moniker of 'Silicon Valley North' is not without justification – see Box 1 for a more detailed comparison on Ottawa-Gatineau and San José. The remainder of the Canadian contingent in the top 30 is made up of Calgary (ranked 22nd) and Toronto (ranked 30th).

Comparing the top 30 Canadian and US cities provides us with a picture of the top end of the distribution of S&E employment and employment shares, but it does not provide an overall picture of how Canadian and US cities compare or how shares of S&E employment vary by city size.

Science and Engineering Share of Employment by City Population Size

Large cities may be particularly advantaged as locations of science and engineering employment. This advantage stems from factors that influence the demand for scientists and engineers by larger urban economies and the supply of scientists and engineers to them. Although it is often difficult to disentangle demand and supply factors, we address them separately below.

The supply of scientists and engineers may be influenced by two forces. The first force relates to the difficulty of matching the highly specialised skills of scientists and engineers with the skill requirements of employers. Large labour markets help to improve the efficiency of this matching process and the quality of the matches because they offer a large set of employers and potential workers (Rosenthal and Strange, 2004). Improving the efficiency of the matching process potentially reduces the search costs of employers and the period of unemployment for workers. Improving the quality of matches increases the productivity of workers, possibly resulting in higher worker wages. Higher wages combined with potentially shorter periods of unemployment raises the incentive for scientists and engineers to move to larger cities.

The second force concerns the incentives that large urban labour markets create for workers to invest in their level of human capital. Large cities have a large number of firms that, in turn, create demand for a wide variety of specialised skills. This creates an incentive for workers to invest in their human capital because there is more likely to

10 San José includes Santa Clara County that forms the heart of 'Silicon Valley.' It does not, however, take into account all of Silicon Valley, which typically includes parts of the San Francisco MSA.

be a close match between their skills and the needs of local firms.[11] In other words, the extra investment that workers make in specialised training is more likely to result in a higher paying job if the city is larger. For workers in a smaller city, there is less incentive to invest in these specialised skills because there are fewer local firms that would require them, reducing the possibility of a good match. With imperfect information about job opportunities in other cities, combined with the economic and social costs of moving, it is likely that fewer students are to invest in a specialised set of skills in a small city compared to a large one. Hence, irrespective of their ability to attract highly skilled workers, large cities may be more effective than smaller ones in generating a supply of specialised forms of human capital, like scientists and engineers, that their employers demand.

On the demand side, the benefits of more efficient and effective labour market matching that helps to attract scientists and engineers may also attract potential employers. Employers in larger labour markets tend to have lower search costs and find workers with a better match to their needs. Both of these may draw employers to larger urban economies. Larger urban areas may also have a higher demand for scientists and engineers because of their industrial compositions. Industries that tend to demand the skills of scientists and engineers may also tend to locate in larger centres. For instance, Information and Communication Technology (ICT) industries that drove the so-called 'New Economy' during the 1990s are heavily concentrated in large cities (Beckstead *et al.*, 2003).

Although we cannot discount the possibility that high-technology industries are attracted to larger cities because of their large pools of highly skilled labour, it is also likely that they are attracted to these centres because their highly diverse industrial structures help to facilitate the transmission of new ideas across firms (Jacobs, 1969) and allow firms to access a wide variety of inputs in a stage of their life cycle when there is considerable uncertainty about which inputs they require (Duranton and Puga, 2001).

We do not attempt to identify the relative significance of these demand and supply factors here. Rather, the point to be drawn is that there are reasons to believe both the demand for and the supply of scientists and engineers may be higher in large cities, and therefore, we should consider population size when we are comparing cities.

---

11 This argument is derived from Kim (1989), whose model illustrates why market size might be positively associated with larger investments in human capital.

To test the relationship between the location of S&E occupations and city size, we divide cities into six size classes. Within each size class, there is at least one Canadian city. The largest size class (cities of 4 million persons or more) includes only 10 cities, while the smallest size class (cities between 100,000 and 249,999 persons) includes some 142 cities. The proportion of workers in S&E occupations is averaged across cities within each size class. To further illustrate the relationship between size and S&E employment shares, Figure 1 also includes the S&E share of employment for Small Urban-Rural areas across Canada and the United States.[12]

Figure 1. Science and engineering average employment shares by size class, 2000 and 2001

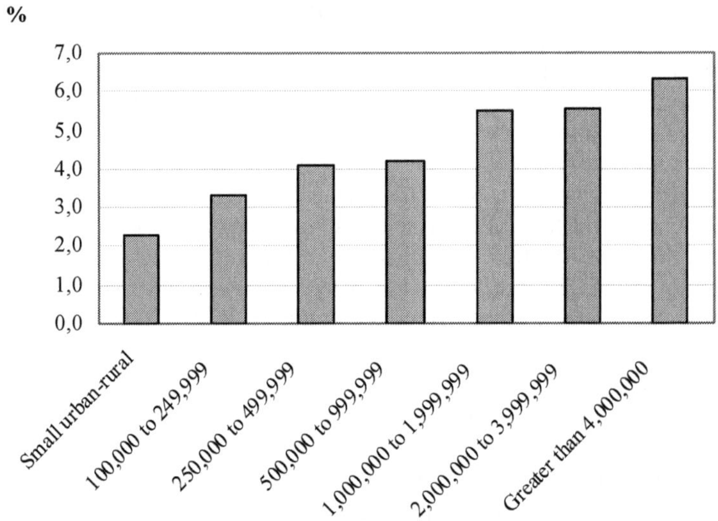

Source: Canadian Census (2001) and US Census (2000).

Figure 1 shows that there is a strong positive association between city size and the share of employment in S&E occupations. For instance, cities with a population greater than 4 million have almost twice the S&E employment share as cities with a population between 100,000 and 249,999 and almost 3 times the S&E employment share of the Small

12 Note that for the Small Urban-Rural category, the S&E employment share is a weighted average of the S&E shares for all urban and rural areas that fall into this category.

Urban-Rural class. Further statistical testing confirms the positive association between city size and S&E employment shares.[13]

One of the implications to be drawn from Figure 1 is that we should be comparing Canadian and US cities of similar size in order to see how Canadian cities rank relative to their US counterparts. Table 6 presents Canadian and US cities for each of the six city-size classes, with each city ranked by its S&E employment share. Since there are 316 cities in our data set, it is difficult to present all of them in this table (see Beckstead and Brown, 2006). Therefore, for each city-size class we report the top three cities, the median city and the bottom city. We also report all Canadian cities that fall within each city-size class.

Table 6. Science and Engineering (S&E) Employment Shares of Total Employment by City Size Class, 2000 and 2001

City size class	S&E share (%)	Rank	
Greater than 4,000,000			
Washington, DC/MD/VA	10.9	1	
San Francisco-Oakland-Vallejo, Calif.	8	2	
Toronto	*6.7*	*3*	
Detroit, Mich.	6.1	5	Median
Los Angeles-Long Beach, Calif.	4.5	10	
Average	6.3	...	
2,000,000 to 3,999,999			
Seattle-Everett, Wash.	8.6	1	
Boston, Mass.	8.3	2	
Denver-Boulder-Longmont, Colo.	7.8	3	
Montréal	*5.5*	*8*	Median
Miami-Hialeah, Flo.	2.5	15	
Average	5.6	...	
1,000,000 to 1,999,999			
San José, Calif.	15.7	1	
Ottawa-Gatineau	*11.6*	*2*	
Raleigh-Durham, N.C.	10	3	
Vancouver	*5.2*	*12*	
Salt Lake City-Ogden, Utah	5	14	Median

13 To test the association between S&E employment shares and city populations illustrated in Figure 1, we estimated S&E share= $\alpha+\beta\ln$ (*population*) + ε using the S&E employment shares (S&E share) and populations of all 316 Canadian and US cities as observations. The model shows there is a positive and statistically significant relationship between population and S&E shares across cities (β=0.008, (robust) se = 0.001, t = 7.12).

Las Vegas, Nev.	2.2	28	
Average	5.5	...	
500,000 to 999,999			
Colorado Springs, Colo.	7.8	1	
Calgary	*7.6*	*2*	
Québec	*6.2*	*3*	
Edmonton	*4.5*	*17*	
Akron, Ohio	4.1	21	Median
Hamilton	*3.9*	*24*	
Winnipeg	*3.8*	*26*	
McAllen-Edinburg-Pharr-Mission, Tex.	1.2	43	
Average	4.2	...	
250,000 to 499,999			
Huntsville, Ala.	11.1	1	
Ann Arbor, Mich.	9.2	2	
Trenton, N.J.	8	3	
Victoria	*5.2*	*17*	
Kitchener	*4.9*	*22*	
Halifax	*4.8*	*23*	
Oshawa	*4.6*	*25*	
London	*4*	*34*	
Windsor	*3.8*	*39*	Median
St. Catharines-Niagara	*2.3*	*66*	
Visalia-Tulare-Porterville, Calif.	1	78	
Average	4.1	...	
100,000 to 249,999			
Nashua, N.H.	11.1	1	
Rochester, Minn.	9.6	2	
Santa Fe, N.Mex.	8.9	3	
St. John's	*5.1*	*22*	
Guelph	*4.9*	*24*	
Regina	*4.8*	*27*	
Saint John	*4.1*	*35*	
Kingston	*4*	*36*	
Moncton	*4*	*37*	
Saskatoon	*3.7*	*42*	
Roanoke, Va.	3.2	57	Median
Sherbrooke	*3.1*	*63*	
Barrie	*3*	*64*	
Trois-Rivières	*2.9*	*70*	
Saguenay	*2.9*	*71*	

Greater Sudbury	*2.8*	*78*	
Thunder Bay	*2.6*	*81*	
Kelowna	*2.6*	*84*	
Chatham-Kent	*2.2*	*98*	
Abbotsford	*1.8*	*116*	
Cape Breton	*1.5*	*132*	
Danville, Va.	0.6	142	
Average	3.3	...	
... not applicable			

Source: Canadian Census (2001) and US Census (2000).

Within each size class many Canadian cities perform well. For instance, Toronto is 3rd out of 10 cities with a population greater than 4 million. Montréal is in the middle (median) of 15 cities in the 2 to 4 million class. Ottawa-Gatineau is ranked 2nd and Vancouver is ranked 12th in the 28 cities in the 1 to 2 million class. In the 0.5 to 1 million size class, Calgary is 2nd and Québec 3rd. Victoria (17), Kitchener (22) and Halifax (23) are the top Canadian cities in the 78 cities in the 250,000 to 499,999 size class, while St. John's (22), Guelph (24) and Regina (27) are the highest ranked Canadian cities of the 142 cities in the 100,000 to 249,999 size class.

It is difficult to tell directly from Table 6 whether Canadian cities fare any better or worse than US cities after controlling for size. For all but the smallest size class, there are more Canadian cities at or above the median than below the median. However, with so few observations for the larger size classes, it is difficult to tell whether these differences are due to chance. Figure 2 plots the relationship between population size and S&E shares for Canadian and US cities. Also included in the graph are two trend lines, one for US cities and a second for Canadian cities. Both trend lines indicate a positive relationship between city size and S&E shares, with the Canadian city line being more positively sloped than the US line.

Although this suggests larger Canadian cities tend to have larger shares of employment in S&E occupations, there is no statistically significant difference between the two trend lines.[14] Hence, Canadian

14 We test whether the coefficients for the Canadian and US trend (regression) lines are different by using the following pooled model: S&E Share = $\alpha + \lambda$Canada + βln (Population) + δln(Population) x Canada + ε, where Canada is a binary variable that takes on a value of 1 if the city is in Canada and 0 if it is in the United States. We are specifically interested in testing the joint hypothesis that $\lambda = 0$ and $\delta = 0$. Using a Wald test that takes advantage of the heteroscedasticity corrected variance-covariance matrix of the estimators, we cannot reject the null hypothesis that both

cities regardless of their size are not (statistically) different than their US counterparts. In short, Canadian cities in 2000 and 2001 were able to match US cities in their ability to employ scientists and engineers.

Figure 2. Science and Engineering (S&E) Employment Share as a Function of ciTy Population Size, 2000 and 2001

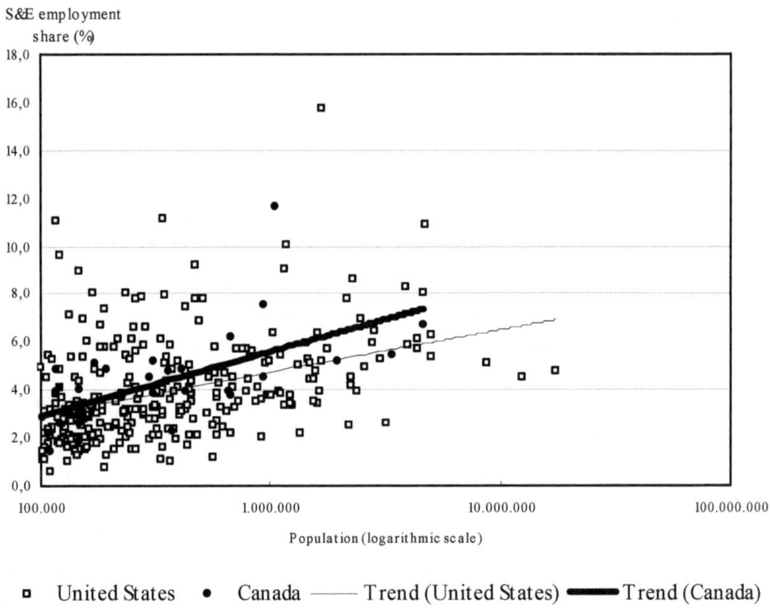

Source: Canadian Census (2001) and US Census (2000).

Growth in Science and Engineering Shares between the 1980 and 1981 Period and the 2000 and 2001 Period

The previous section provided a 'snapshot' of how Canadian cities compared to US cities in 2001. However, it did not give us a picture of how Canadian cities have evolved over time in comparison to their US counterparts. In this section, we provide such a picture. We also go a step further and ask whether S&E employment shares grew faster in larger cities than in smaller cities and in Canadian cities compared to their US counterparts, irrespective of their size.

coefficients are equal to zero (F = 1.16, Prob > F = 0.31). That is, the coefficients that underlie both trend lines (their intercepts and slopes) are not statistically different from each other.

To facilitate the comparison of Canadian to US cities, we utilise a subset of cities that had a population of 100,000 or more in both 1980-1981 and 2000-2001. There were a total of 264 cities across both countries that met this criterion. As above, it is very difficult to present information on all cities. For the interested reader, these are available in a separate paper.[15]

Comparing the average S&E employment shares of Canadian and US cities within this subset reveals stronger growth for Canadian cities. In 1980 and 1981, the average S&E employment share for US cities was 2.4%, while for Canadian cities it was 2.6%. By 2000 and 2001, the average US city share had risen to 4.0%, while the average Canadian city share had risen even more to 4.6% (Figure 3). While changes in average S&E shares of US and Canadian cities suggest broad trends, they do not provide us details regarding the individual performance of Canadian cities.

Figure 3. Average Science and Engineering Employment Share Change for Canadian and US Cities: 1980, 1981 and 2000, 2001

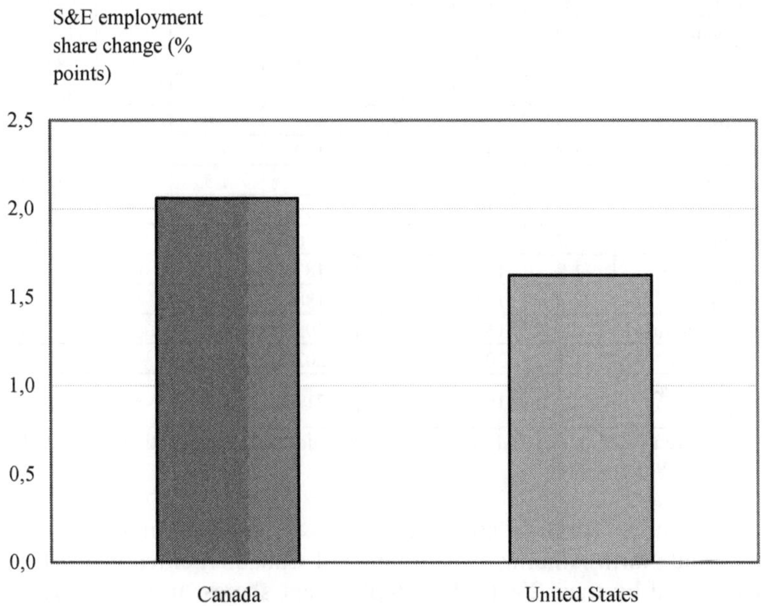

Source: Canadian censuses (1981 and 2001) and US censuses (1980 and 2000).

15 See Beckstead and Brown (2006).

Table 7. Science and Engineering Employment Shares across Canadian Cities, for the Years 1980, 1981 and 2000, 2001

	1981		2001		Difference between 1981 and 2001		2001 Population
	Share (%)	Rank*	Share (%)	Rank*	Share (%)	Rank*	
Ottawa-Gatineau	5.2	8	11.6	2	6.4	6	1,063,664
Calgary	5	9	7.6	17	2.6	-8	951,395
Toronto	3.2	51	6.7	25	3.5	26	4,682,898
Québec	3.5	34	6.2	32	2.7	2	682,755
Montréal	2.6	96	5.5	53	2.9	43	3,426,350
Victoria	2.8	81	5.2	64	2.4	17	311,924
Vancouver	2.5	106	5.2	67	2.6	39	1,986,965
St. John's	2.7	88	5.1	69	2.4	19	172,928
Kitchener	2.3	132	4.9	80	2.6	52	414,284
Regina	2.6	101	4.8	82	2.3	19	192,802
Halifax	2.5	102	4.8	85	2.3	17	359,183
Oshawa	2.3	134	4.6	91	2.3	43	296,298
Edmonton	3	67	4.5	95	1.6	-28	937,840
Saint John	1.9	168	4.1	109	2.2	59	122,660
Kingston	2.5	107	4	113	1.5	-6	146,838
London	2.1	151	4	119	1.8	32	432,452
Hamilton	2.3	124	3.9	120	1.6	4	662,401
Windsor	1.8	187	3.8	129	2.1	58	307,877
Winnipeg	2.3	131	3.8	132	1.5	-1	671,275
Saskatoon	2.4	119	3.7	139	1.3	-20	225,928
Sherbrooke	1.7	190	3.1	176	1.4	14	153,821
Trois-Rivières	1.8	184	2.9	187	1.2	-3	137,507
Saguenay	2.3	125	2.9	188	0.6	-63	154,937
Greater Sudbury	2	163	2.8	195	0.8	-32	155,601
Thunder Bay	1.7	200	2.6	202	1	-2	121,985
St. Catharines-Niagara	2	159	2.3	219	0.3	-60	377,009

* Cities are ranked out of 264 North American cities with a population of 100,000 or more in 1981 and 2001.

Meanwhile, Table 7 compares the S&E occupation employment shares and North American share ranks of Canadian cities in 1981 and 2001, sorted by their 2001 S&E employment shares. In both 1981 and 2001, Ottawa-Gatineau and Calgary topped the list in Canada in terms of their employment shares. In North American terms, Ottawa-Gatineau moved its rank from 8th to 2nd, while Calgary's rank fell from 9th to 17th despite its above-average S&E share increase. Remarkably, Ottawa-

Gatineau increased its share by 6.5 percentage points. This was the second highest percentage point increase after San José (see Box 1 for a more extensive comparison of Ottawa-Gatineau and San José) and well above the average city increase of 1.7 percentage points.

Several cities have increased their rank substantially. These cities include Canada's largest cities – Toronto, Montréal and Vancouver. Toronto increased its rank by 26 places (51st to 25th), Montréal increased it rank by 43 places (96th to 53rd) and Vancouver increased its rank by 39 places (106th to 67th). Other notable increases in ranks were Saint John (168th to 109th), Windsor (187th to 129th) and Kitchener (132nd to 80th). Although all Canadian cities increased their share of S&E employment, not all cities fared well relative to their North American counterparts. In particular, the Saguenay and St. Catharines-Niagara's ranks fell by 63 and 60 places, respectively.

The previous section demonstrated a strong positive relationship between the size of a city and its S&E share of employment. A cursory review of Table 7 suggests this may also be the case for S&E share growth. The cities with the largest populations tended to have the largest increases in the S&E employment shares.

There are several reasons why we might expect S&E employment share growth to be strongest in larger cities. First, if over time S&E workers have invested in more specialised skills, then more of these workers may be attracted to the larger labour markets that characterise more populous cities (Kim, 1989). It may also be that as the prevalence of workers with very specialised occupations increases over time so too do the shares of couples where both partners have specialised skill sets. Satisfying the career goals of both partners – which effectively exacerbates the matching problem – would tend to further increase the attractiveness of large labour markets to skilled workers like scientists and engineers (Costa and Kahn, 2000). Finally, if industries that have a high demand for S&E workers are growing more rapidly in larger centres, then S&E employment growth may again be stronger in larger centres (Beckstread *et al.*, 2003).

We can test the relationship between population size and S&E employment share change by plotting S&E employment share change against population levels of all US and Canadian cities in 1981 (see Figure 4). We do so by taking the average of the S&E employment share change of cities that fall into the six city size classes used in Figure 1.

Figure 4. Average Science and Engineering Share Change by 1980 and 1981 Population Size Class

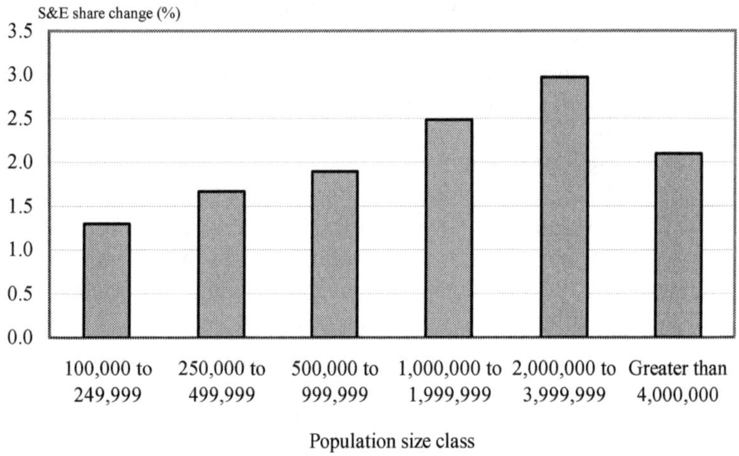

Source: Canadian Census (1981) and US Census (1980).

There is a clear positive relationship across the first five size classes between 1980-1981 population size and subsequent increases in S&E employment shares (Figure 4). Cities with a population greater than four million on average experienced a smaller increase in their S&E employment share than cities with a population between one and two million. The relatively poor performance of this more populous group was largely due to Los Angeles. With Los Angeles excluded, the average share increase of this city size class was 2.8%. This is essentially the same as the second largest city size class and does suggest there is some levelling off of the effect of city size (which may be due to the initially high shares of S&E employment in these cities). More extensive statistical testing confirms the positive association between city size and S&E share change.[16]

[16] To formally test the independent effect of population levels and Canada on the percentage point change in S&E shares ($\Delta S\&E$) across cities, we estimated the following model:

$$\Delta S\&E = \alpha + \beta S\&E1980 + \delta \ln(1980 \text{ Population}) + \gamma \text{Canada} + \varepsilon,$$

where S&E1980 is the S&E share in 1980 and 1981 for a city *i*, 1980 Population is the population of the city in 1980 and 1981 and Canada is a binary variable that takes on a value of 1 if the city is in Canada and 0 if it is located in the United States. The initial S&E share is included in the model to control for the fact that, by its very nature, percentage point share change will tend to be higher in cities with initially high

This chapter focuses on how Canadian cities compare to US cities. Given our sample of cities, the average Canadian city is 85% the size of the average US city, one might expect that their increase in S&E employment shares would be less than US cities. As Figure 3 (above) demonstrates, the change in S&E employment share for Canadian cities, irrespective of their size, was stronger than US cities. This is confirmed through more extensive statistical testing. After controlling for the size of cities, and their initial S&E employment share, the difference in the share change for Canadian cities was statistically significant.[17] The broad conclusion to be taken from this section is that not only are larger cities advantaged in their ability to support high shares of S&E employment, but also this advantage appears to be increasing over time. It also appears that S&E employment shares have grown more rapidly in Canadian cities.

Silicon Valley: North versus South

Ottawa-Gatineau has often been given the moniker 'Silicon Valley North' in an obvious reference to the original 'Silicon Valley,' which is located primarily in the San José metropolitan area at the southern end of San Francisco Bay. The evidence presented elsewhere in this chapter suggests there may well be some merit to this comparison. In this sec-

S&E shares. To see why, consider an instance where looking across cities there is no difference in their rate of S&E employment growth. If S&E employment were to double, while all other forms of employment remained constant, a city with an initial S&E employment share of 1% would increase its S&E share by approximately 1 percentage point while a city with an initial share of 5% would raise its S&E share by approximately 5 percentage points. It is important to note that the coefficient on S&E1980 does not provide us with a test of whether cities with an initially high S&E shares experience above- or below- average growth – divergence from, or regression to, the mean. If cities are experiencing a process of divergence from the mean or, up to a point, regression to the mean, the coefficient on S&E1980 will tend to be positive. Turning to the model's results, as expected, we find a positive and statistically significant relationship between the initial S&E share of a city (β=0.35, (robust) se = 0.14, t = 2.36) and its S&E employment share change. However, we also find an independent, positive and statistically significant effect of initial population size on S&E share change (δ=0.0032, (robust) se = 0.00096, t = 3.31). Finally, the coefficient on the Canada variable is positive and statistically significant (λ=0.0041, (robust) se = 0.0019, t = 2.21). Hence, Canadian cities, after controlling for their initial S&E share and their population size, experienced a faster increase in their S&E share than US cities.

17 As we note in Footnote 11, it is unlikely that the difference we observe between US and Canadian cities is due to chance. Moreover, the model demonstrates that Canadian cities experienced a significantly greater increase in their S&E shares after controlling for their population levels and the initial size of their S&E share of employment.

tion, we develop a more comprehensive, but by no means complete, comparison of these two high-technology centres. Table 8 provides a comparison of the occupational characteristics of Ottawa-Gatineau and San José.

Table 8. Ottawa-Gatineau and San José Science and Engineering (S&E) Employment Comparison

	Ottawa-Gatineau				San José			
	1981		**2001**		**1980**		**2000**	
Population	711,920		1,063,665		1,297,507		1,688,089	
Employment	360,050		561,871		662,452		847,345	
	Employ-ment	Share	Employ-ment	Share	Employ-ment	Share	Employ-ment	Share
S&E employment	18,720	100	65,330	100	52,273	100	133,121	100
Business sector	7,330	39	43,990	67	49,070	94	125,710	94
Non-business sector	11,395	61	21,340	33	3,203	6	7,411	6
Computer and information scientists	6,945	37	34,960	54	13,528	26	36,400	27
Life and related scientists	1,045	6	1,080	2	600	1	2,707	2
Physical and related scientists	1,355	7	1,980	3	2,500	5	5,544	4
Social and related scientists	3,470	19	7,185	11	2,242	4	3,343	3
Engineers	5,910	32	20,120	31	33,403	64	85,127	64
Information technology professionals	...	...	46,562	71	...	...	79,764	60
S&E share of employment (%)	5.2	...	11.6	...	7.9	...	15.7	...
... not applicable								

Note: Share is in percentage of science and engineering employment.

Source: Canadian censuses (1981 and 2001) and US censuses (1980 and 2000).

Ottawa-Gatineau's population is smaller than San José's, but Ottawa-Gatineau's population growth has been stronger. In 1980 and 1981, Ottawa-Gatineau's population was about half of San José's. But between the 1980 and 1981 period and the 2000 and 2001 period, Ottawa-Gatineau's population grew more rapidly. As a result, Ottawa-Gatineau's population was about two-thirds of San José's by 2000-2001. S&E employment follows a similar pattern. Ottawa-Gatineau's S&E employment was 29% of San José's in 1980 and 1981, but by

2000 and 2001 this proportion had increased to 40%, reflecting faster S&E employment growth in Ottawa-Gatineau over the intervening two decades.

Dividing S&E employment by the number of workers employed in the business and non-business sectors, reveals some important differences between San José and Ottawa-Gatineau. Reflecting Ottawa-Gatineau's status as a national capital, a much higher proportion of its science and engineering population was in the non-business sector. In 1980 and 1981, 61% of Ottawa-Gatineau's S&E workforce was in the non-business sector, while this sector accounted for only 6% of San José's S&E workforce. Although, by 2001, Ottawa-Gatineau's S&E workforce had undergone a remarkable transformation towards a stronger private sector component with two-thirds of its workforce in the business sector. Meanwhile, the business sector continues to be a much stronger force in maintaining San José's S&E workforce.

If S&E employment is divided into its component occupations, further differences emerge. More of Ottawa-Gatineau's S&E workforce was made up of Social and Related Scientists – 11% in Ottawa-Gatineau versus 3% in San José in 2000 and 2001. Ottawa-Gatineau's employment has also been increasingly dominated by Computer and Information Scientists (54% of S&E employment in 2001) while San José's employment was dominated in both 1980 and 2000 by Engineers (64% of S&E employment in both years). This does not necessarily mean, however, that San José was less oriented towards information technology than Ottawa-Gatineau given that 45% of San José's Engineers were computer software engineers in 2000. A more direct comparison can be made by combining Computer and Information Scientists with Software Engineers (and related computer hardware engineering occupations) into a separate classification – what we could call Information Technology Professionals.

In 2001, Ottawa-Gatineau had about 46,600 of these Information Technology Professionals that accounted for 71% of Ottawa-Gatineau's S&E employment while, in 2000, San José had 79,800 Information Technology Professionals, which accounted for 60% of San José's S&E employment. San José's Information Technology Professionals workforce was larger, but accounted for a smaller share of its S&E employment than Ottawa-Gatineau. At the zenith of the high-technology boom, Ottawa-Gatineau's science and engineering workforce was more oriented towards information technology than San José.

Conclusions

In recent years, cities have become increasingly interested in their ability to increase their level of human capital. In this chapter, we use employment in science- and engineering-based occupations as a measure of human capital and ask how Canadian cities compare to their US counterparts. We chose to compare Canadian to US cities not only because the United States is often used as a benchmark for Canada, but also because the market for skilled labour is at least continental in scale. This holds not only for science and engineering workers born in North America, but also for those thinking of migrating to North America from abroad.

We find that science and engineering employment is not spread equally across cities. For some cities, 1 in 10 employed workers is in science- and engineering-based occupations, while in others, this ratio falls to only 1 in 100. Yet, whether one city does better than another is not purely random. There is a strong positive association between city size and the share of its workforce in science and engineering occupations and the growth of this share over time. Big cities do better than small cities. Not only are larger cities advantaged, but their advantage has been increasing. On average, larger cities experienced a greater increase in their S&E employment share between the 1980 and 1981 period and the 2000 and 2001 period.

In 1980-1981, the science and engineering share of employment in Canadian cities did lag slightly behind US cities. But in the ensuing two decades, science and engineering employment in Canadian cities grew at a faster pace than in US cities. As a result, the science and engineering base of Canadian cities was higher than US cities by the end of the period, albeit on a city by city basis; there is no statistically significant difference. At the very least, these results suggest Canadian cities have held their own.

Although we provide some explanations as to why larger cities might be advantaged, the focus of this chapter was not on trying to test why population size matters or whether other factors were important. A particularly important avenue for research is the question of what factors affect the growth in science and engineering employment. How important is the industrial make-up of a city? How important is the presence of universities? Does a heavy governmental presence help or hinder growth? Do factors that make cities attractive – e.g., cultural activities, climate, low-crime rates – also influence growth in science and engineering employment? The research presented in this chapter provides important insights into how knowledge cities are and can be characterised according to employment profile. At the same time, these latter questions represent how future research can be used to establish causation.

References

Atkinson, R.D., Court, R.H. (1998). *The new economy index: understanding America's economic transformation.* Retrieved July 20, 2006, from www.neweconomyindex.org/ section3_page15.html.

Beckstead, D., Brown, W.M. (2006). Innovation capabilities: comparing science and engineering employment in Canadian and US cities. *Statistics Canada.* Catalogue No.11-622-MIE2006012. Ottawa: Statistics Canada.

Beckstead, D., Brown, M., Gellatly, G., Seaborne, C. (2003). A decade of growth: the emerging geography of new economy industries in the 1990s. *Statistics Canada.* Catalogue No.11-622-MIE2003003. Ottawa: Statistics Canada.

Beckstead, D., Gellatly, G. (2006). *Innovation capabilities: science and engineering employment in Canada and the United States.* The Canadian Economy in Transition Research Paper Series, Catalogue No.11-622-MIE2006. Ottawa: Statistics Canada.

Costa, D.L., Kahn, M.E. (2000). Power couples: changes in the locational choice of the college educated, 1940-1999. *The Quarterly Journal of Economics, 115* (4), 1287-1315.

DeVoretz, D.J., Coulombe, D. (2005). Labour mobility between Canada and the United States: quo vadis 2004? In *Social and Labour Market Aspects of North American Linkages.* R.G. Harris and T. Lemieux (eds.). Calgary: University of Calgary Press. 417-443.

Duranton, G., Puga, D. (2001). Nursery cities: urban diversity, process innovation, and the life cycle of products. *American Economic Review, 91* (5), 1454-1477.

Florida, R. (2002a). The economic geography of talent. *Annals of the Association of American Geographers, 92* (4), 743-755.

Florida, R. (2002b). *The rise of the creative class: and how it's transforming work, leisure, community and everyday life.* New York: Basic Books.

Gertler, M.S., Florida, R., Gates, G., Vinodrai, T. (2002). *Competing on creativity: placing Ontario's cities in North American context.* Toronto: Ontario Ministry of Enterprise, Opportunity and Innovation and the Institute for Competitiveness and Prosperity.

Glaeser, E.L., Saiz, E. (2004). *The rise of the skilled city.* NBER Working Paper Series. Working Paper 10191. Cambridge, MA: National Bureau of Economic Research.

Gu, W., Kaci, M., Maynard, J-P., Sillama, M-A. (2003). The changing composition of the Canadian workforce and its impact on productivity growth. In *Productivity Growth in Canada – 2002.* Catalogue No.15-204-XPE. J.R. Baldwin and T.M. Harchaoui (eds.). Ottawa: Statistics Canada.

Jacobs, J. (1969). *The economy of cities*. New York: Random House.

Kim, S. (1989). Labor specialization and the extent of the market. *Journal of Political Economy*, *97* (3), 692-705.

Lucas, R.E., Jr. (1988). On the mechanics of economic development. *Journal of Monetary Economics*, *22*, 3-42.

National Science Board. (2004). *Science and Engineering Indicators 2004, Volume 1 NSB 04-1*. Arlington, VA: National Science Foundation.

Polèse, M., Tremblay, R. (2005). Le classement des villes dans l'économie du savoir: une analyse intégrée des régions urbaines Canadiennes et Américaines. *The Canadian Geographer*, *49* (2), 198-213.

Rosenthal, S., Strange, W. (2004). Evidence on the nature and sources of agglomeration economies. In *Handbook of Urban and Regional Economics*. J.V. Henderson and J.-F. Thisse (eds.). Amsterdam: Elsevier.

CHAPTER 9

Reflexive Governance and City-Region Economic Change

Between Reflexivity and State Selectivity in the Governance of Ottawa's Economic Transition

Joseph LEIBOVITZ

Institute of Geography,
University of Edinburgh

Introduction

With its emphasis on the mediation of state and market actors, the complex nature of policymaking in increasingly networked societies, and the significance of trust and reciprocity in social and economic life, an institutional perspective can offer considerable understanding to the qualitative aspects of local economic transformations (Amin, 1999). This chapter highlights how communities anticipate, adapt and respond to economic challenges and structural transformation. The potential contribution of institutionalist perspectives on city-region economic change is therefore in drawing attention to how economic forces and technological change are mediated by actors (Cooke *et al.*, 1998).

In his seminal work on the history of the city of Ottawa, Taylor has argued that despite the rapid transformation of Ottawa since 1945, conspicuous forces of continuity in the political economy of the city and its region can still be detected:

> …much remains as it was. Fragmentation, marginality, and corporate oversight are still central. Though the main city is probably less divided than in any time in its history, a new sort of division, always latent, and situated at the regional level, has emerged strongly since the war. Large, territorially-based, self conscious communities compete with Ottawa proper for both business and influence…large regional governments contend with each other and within themselves…There is a sense of population overweight,

> and probably the major reason is that there is still no common agenda, and still no community able to set one. It is a metropolis made up of undigested pieces, reluctant or unable to co-operate. The sense of agglomeration of small communities persists (1986, p. 209).

Without resorting to historical determinism, the more recent transformation of Ottawa, from a city known for the dominance of federal government employment – in a sense a single industry town – to a hotbed of entrepreneurial and high-technology activities, tends to resemble some of the properties discussed by Taylor. The story of institutional change in Ottawa and the role that institutions have played in coaxing this transformation reveal how economic change is always embedded in a wider political economy (Agnew, 2000).

The action of regional actors underpins the role that power and interests play in underwriting the public atmosphere and legitimacy within which change takes places. The growth of the high-technology cluster in Ottawa has been accompanied by implicit and explicit forms of re-ordering social relations and institutional orientation. Ottawa has exhibited a remarkable constellation of institutional networking at different levels, from federal through provincial to local levels of activity. While it may be impossible to capture the entire complexity of this process, this chapter aims to demonstrate how economic change has been mitigated by and interwoven with a shift in public discourse, state intervention (explicit and implicit), and the construction of legitimacy around the centrality of high technology to the prosperity of the region.

Reflexivity and Adaptation
Institutions and the Mediation of Territorial Economies

For some time now, economic sociologists and geographers have been arguing that the economic success of cities, regions and nations is not merely a product of 'hard' economic factors, as important as they are. Rather, it has been suggested that collective action, inter-personal and institutional ties, trust, collaboration, social behaviour, and norms and routines, are all elements which also underpin and shape the expectations and conduct of economic actors (Amin and Thrift, 1994; Putnam, 1993; Martin, 2000; Piore and Sabel, 1984).

A particular strand of this institutionalist view of the economy focuses on the governance structures through which economies are managed. In relation to the challenges of regional economic performance, it has been argued that associative forms of governance can enhance the competitiveness of territorial economies (Amin, 1999; Amin and Thrift, 1995; Cooke and Morgan, 1998; Gertler, 2001). Associative governance involves policy-making processes and institutional infrastructures that facilitate the participation of and interaction between a wide range of actors from public agencies, private enterprises, trade unions, and the voluntary sector. Such processes are said to increase the capacity of communities in developing inclusive, responsive and strategic frameworks of economic governance. In short, associative economic governance is an important institutional vehicle for the promotion of 'progressive competitiveness.'

The attractiveness of institutionalist perspectives on urban and regional development has been in drawing attention to the embeddedness of economic life in the wider spheres of social, political, and cultural worlds. Innovation, for instance, has been discussed as socially embedded in processes of collective and interactive learning and adaptation (Lundvall and Maskell, 2000). The strategy and choices of firms are also conditioned by constraints and opportunities represented by previous experience, social expectations, the structure of capital markets, and the collective norms and conventions held by economic actors (Hodgson, 1993). And the emergence, growth and possible decline of firms and clusters of industrial activities are also an outcome of the interaction of individual choices as well as an evolutionary and path-dependent dynamics, where historical legacies and chance discovery play an important role (Arthur, 1994).

There are important geographic and regional governance implications to these arguments, although these implications have not always been made as explicit as they should have been in the literature on urban and regional politics (see Cumbers *et al.*, 2003; Swyngedouw, 2000). In particular, economic geographers have argued that innovative and technological development capacities are necessarily spatially differentiated by the ability of certain actors and institutions, within particular locales, to engage in collaborative learning and collective strategic anticipation of market opportunities (Maskell *et al.*, 1998; Saxenian, 1994). These capabilities, in turn, enable certain regions to keep to the 'high road' of economic competitiveness based on the ability to respond rapidly to volatile market conditions (Cooke, 1995). Furthermore, flows of knowledge and the nurturing of networking, trust and collaboration are enhanced by territorial proximity where face-to-face interaction and even cultural attributes are said to play a persistently significant role in the successful development and use of advanced technologies (Gertler and Wolfe, 1998; Storper, 1997).

The continued strength of geographical factors in underpinning economies places a premium on regional or city-region scales of economic and political organisation (Parr, 2005). The appearance of effective governance frameworks cannot come 'out of nowhere.' It is deeply embedded in the political economy in which actors are situated, in local and regional histories, and in certain contingencies. In fact, critical approaches to studies of institutions and urban and regional economic governance have recently drawn attention to the wider political economy in which these are situated, pointing to the centrality of state strategy, power relations, and ideological stances in shaping particular institutional forms (Amin, 2004; Cumbers *et al.*, 2003; Keating, 2001; Leitner and Sheppard, 2002; Lovering, 1999; MacLeod, 2001). The path-dependent nature of these institutional features of economic life constitutes the core of the following discussion.

'The End of Ottawa'? The Rise of Enterprise Discourses in the Late 1990s

Despite the fact that Ottawa's high-technology cluster is not the dominant sector within the structure of the region's economy, it has displayed impressive growth in the number of firms and employees, and has helped to generate a changing regional image and identity (Mallett, 2002; Steed and DeGenova, 1983). As Andrew's citation of the early 1980s optimism and excitement suggests, the emergence of 'Silicon Valley North' as a dominant image did go a long way in implying a regional cultural transformation:

> An infectious boom-town atmosphere is gripping Ottawa these days – the kind that has not been felt here since Queen Victoria chose the city as Canada's capital more than 100 years ago. Ottawa is no longer a slightly sleepy, one-industry town. Red tape is being smothered in computer age; bureaucrats are watching anxiously as engineers move in next door; traffic on the Queensway flowing both ways in rush hour. High technology – a whole series of new business made possible by computerisation and miniaturised electronics – has come to town (quoted from the local press, source unspecified, in Andrew, 1983, p. 148-149).

The sense of regional cultural transformation has indeed played a central role in, and has become a key aspect of, the changing institutional milieu of the region.

The Ottawa technology-oriented cluster is skewed towards the telecommunications sector in terms of employment and number of establishments, with test and measurement, computer software design, computer hardware production, and defence-related technology firms following as significant 'clusters' within the overall complex (Ottawa Planning Department, 2005; Steed and DeGenova, 1983). Moreover, the proportions of research and development jobs in the region are significant. According to an employment survey conducted by the Regional Municipality of Ottawa-Carleton, 16,000 jobs (4% of total employment) were classified as R&D in 1999. Of the technology-oriented jobs, 32 % (just above 14,000) were R&D-related, with the telecommunications industry in particular displaying considerable strength in this area, containing 73% of the region's total R&D jobs.

The growth of the technology-oriented cluster in Ottawa has been enabled by the somewhat unique configuration of historical and institutional elements. By the late 1990s, this sector displayed considerable innovative, product development and marketing capacities, primarily through firms such as Newbridge Networks, Corel, Nortel, Cognos, Mitel and JetForm, several of which have secured significant market shares for their product (JetForm, for instance, has an estimated share of 80% of the electronic forms market), have been involved in international strategic alliances and acquisitions (Mitel and Newbridge, for example), and have engaged in fierce competition against global high-technology giants (as exemplified by Corel's attempt to bite into Microsoft's dominance with its adoption of WordPerfect and CorelDraw). Thus, many of the firms in Ottawa's technology-oriented sector can be said to engage in quality-based rather than price-based competition, a conclusion also reached by Steed and DeGenova in the early 1980s.

The presence of federal research capabilities certainly helped initially to carve a comparative advantage for Ottawa in terms of scientific and technological knowledge, and later on in terms of product development, as the commercialisation of technologies became more common, and demand remained rather stable due to a 'secured' market in the form of government procurement (Doyle, 1996). Equally important, however, is the *social* aspect of Ottawa's high-technology growth. As mentioned earlier, the close linkages between the 'technological entrepreneurs' and the traditional resource industries of the Ottawa Valley resulted in a unique marriage between technological know-how and entrepreneurship on the one hand, and management know-how and venture capital financing on the other hand (Harrison *et al.*, 2004).

The relative stability of the Ottawa economy came to an abrupt halt during the middle to late 1990s, when some 20,000 government jobs were about to be lost, heralding – for some regional actors – "the end of Ottawa" (based on interviews during 1999 and 2000). This crisis in Ottawa expressed itself in the increasing salience of discourses and policy actions – espoused by economic governance institutions and prominent figures from industry – which emphasised the virtues of private enterprise and entrepreneurship on one hand, and the imperative of Ottawa to 'liberate' itself from the legacy of 'bureaucratic-minded' public sector-dominated economy on the other hand. As a result, the perceived need to underwrite a transformation in the regional economic culture became one of the most important goals of the institutional apparatus in the National Capital Region. The following perspective represents a commonly held view of the story of economic and cultural change in Ottawa, and how it might be different from other city regions in Canada:

> In recent years an entrepreneurial culture has evolved in Ottawa-Carleton beside the traditional government-town culture. For example, most of the existing high-tech firms were developed by residents rather than being branch plants. This is a unique situation found in only a few major locations, such as Boston and Silicon Valley in California. The entrepreneurial spirit has now entered government research labs in the region, as they have become more focused on practical research and ties to the private sector. More partnerships between governments at all levels and businesses are being formed. Other initiatives, such as improved business links with the local universities and colleges, more sources of venture capital, and better support for small and medium enterprises, are further enhancing the business climate in Ottawa-Carleton (RMOC, 1996, p. 4-5).

Throughout the 1990s, therefore, the realisation that Ottawa's economy is embedded within a wider shift towards a 'new' knowledge-based economy has played a central role in perceptions and attitudes regarding regional economic renewal and Ottawa's place within an emerging spatial division of labour. This realisation was accompanied by considerable institutional alignment between federal institutions and local actors. This constellation of forces has created a distinctive conjuncture of discourses and actions in Ottawa – one that has provided Ottawa with a distinctive milieu for collective enterprise and reflexive action.

Constructing Reflexivity
Ottawa's Institutional Plethora

The following discussion traces the actual process of institutional development and change in the Ottawa region since the mid-1980s, with particular emphasis on the Ontario part of the National Capital Region (NCR). The discussion draws on some 60 interviews conducted with representatives of a wide range of technology-intensive firms, local and regional economic development institutions, and government offices at the local, regional, provincial and federal levels. The specific aim of the discussion is to point to particular *institutional moments* in the development of economic governance in the Ottawa region, *viz.*, to important periods of change in which the complex interplay of political and economic forces operating at various geographic scales worked to produce particular institutional forms in Ottawa.

Institutional Development and Change

By the late 1990s, the complex institutional array operating within the NCR provided a fascinating example of the unique political, social and economic geography of the region. Combining different levels of government agencies, quasi-public authorities, one-off, *ad hoc*, and 'permanent' public-private partnerships, voluntary and not-for-profit organisations, and a plethora of industrial associations, spanning different spatial scales in their apparatus, the map of institutions in the Ottawa region is undoubtedly rich and complex. In quantitative terms, this institutional map suggests that the region does possess considerable *potential* to achieve 'institutional thickness,' providing considerable opportunity for associative economic governance to take place.

In addition to the Ottawa Economic Development Corporation (OED), which was set-up originally as the Ottawa Industrial Development Commission in the early 1960s (CIDC), and the economic development offices of individual municipalities on both sides of the Ottawa River (which were all amalgamated on either side in 2002), the region hosts two federal 'research-rich' institutions, the National Research Council (NRC), and the Communication Research Centre (CRC); the Ottawa Board of Trade (OBT) as well as an array of local chambers of commerce; the Ottawa Tourism and Convention Authority (OTCA); the Regional Economic Diversification Opportunity (REDO) partnership; the economic development corporation of the Québec part of the National Capital Region (Société de diversification économique de l'Outaouais); the Eastern Ontario Division of the Ministry of Economic Development, Trade and Tourism; the federal National Capital Commission (NCC), and a host of collaborative research and development organisations, industrial associations and networking institutions such as the Ottawa Life Sciences Council (OLSC), the Canadian Advanced Technology Association (CATA), and the Ottawa Centre for Research and Innovation (OCRI) and its 'subsidiaries' such as New Media North (a partnership of multi-media and cultural industries), 'Zone 5ive' (an association of technology marketing firms), OCRInet (a not-for-profit partnership promoting linkages between Ottawa's technology-oriented industrial complex, government and educational institutions, and the provision of technological infrastructure), the Telecommunications Research Institute of Ontario (TRIO), PRECARN Associates (a consortium of nearly 40 high-technology organisations, R&D institutes, and federal government agencies which promotes collaborative research relating to Canada's industrial competitiveness), the Optical Processing and Computing Consortium of Canada (OPCOM), PARTNERS (yet another business-education partnership aimed to "foster a superior, cost-effective education system and an internationally successful business community" (OED 1997, p. 36)), and the Industrial Research Assistance Programme (IRAP), managed by the NRC to provide for a technology-transfer network for small and medium enterprises.

The institutional fabric of Ottawa is thus rich and complex, cutting across various state and private sector organisations, and spatial scales of governance. Importantly, this institutional fabric provides a potential (and explicitly intended) platform for 'thick' synergy between the state and the market. Theoretically, such synergy forms the backbone of the associative view of economic governance by providing for an 'idea rich' innovation milieu, supported by intermediary institutions which are inserted between the state and the market, and thus facilitating frequent dialogue, formal and informal collaboration, the sharing of knowledge and ideas, and the formation of community identity.

Indeed, members of the region's high-technology community did point to the existence of shared norms, experiences, and 'community feeling,' which underpinned the emergence of the technology-oriented cluster in Ottawa. In interviews, a variety of factors facilitated the frequent networking and 'talk' between firms and have all been mentioned as factors contributing to the solidification of the kind of untraded interdependencies emphasised by Storper (1997) as important ingredients in regional economic success. Some of these factors include the following: the common origins of high-technology firms in federal government research laboratories and in a network of spin-offs companies, such as Mitel, Newbridge Networks and Corel; the relatively small size of the region (and the high-technology community); the clustering of the region's technology-oriented complex in the west-end of the region; the relative isolation of the NCR from the provincial 'power centres' in Toronto and Québec; and the role of institutions such as OCRI in solidifying the technology-oriented complex's identity and visibility.

As Streeck and Schmitter (1985; quoted in Cooke and Morgan, 1998, p. 22) have argued, institutions of associative governance have an important role to play in combining the distinctive assets of the state and capital, and in providing the necessary co-ordinative mechanism that is required in a multi-stakeholder environment:

> Institutions of this kind, in addition to mediating between the state and the market, seem to limit the extent to which the two can invade each other's expense. In this way, they seem to inject an element of stability in their re-

> spective polities which makes them less subject to changing political fashion. Also, an elaborate intermediary associational structure seems to enlarge a country's repertoire of policy alternatives – its 'requisite variety' – and this may enable such countries to respond to new problems without having to undergo dramatic internal realignments.

Within the particular policy area and institutional plethora revolving around economic governance, the story of the Ottawa Economic Development Corporation (OED) is of particular interest. As an explicit partnership designed to support the growth of local companies, assist in the formation of new businesses, attract inward investment, and (importantly) provide for constant public-private dialogue and a multi-stakeholder forum for reflexive strategic action, the OED provided an interesting example of an experiment designed to provide grass-roots approach to economic governance, based (at least theoretically and rhetorically) on associative principles of 'strategic anticipation.'

However, the institutional tissue of the region comprises many other agencies and associations which collectively comprise its system of industrial governance, and – potentially – its institutionalised basis of 'collective social order' (Scott, 1996; Storper and Scott, 1989). One of the key aspects of this institutional framework has been its support of collaborative research and development. Leading centres of research and development in the region include the two federal agencies – the Communication Research Centre and the National Research Council – as well as the two universities, each employing more than 1,000 R&D-dedicated professionals (OED, 1997). In addition, it is estimated that 1 billion Cdn$ a year is spent on R&D by other high-technology institutions. For example, 24 life sciences-related research institutes in Ottawa invest an estimated 300 million Cdn$ annually in R&D, and the proportion of revenue spent on R&D among many firms exceeds 10%.

Perhaps the leading example of a privately-led R&D intensive environment is provided by Nortel, with its 8,000 workers in the region, the great majority of which are scientists and engineers engaged in a broad range of communications-related research. Since the early 1970s, Nortel has reported persistent expansion of 10% a year in its Ottawa workforce, and in the early 1990s announced its plan to add additional 5,000 communications research positions in its western Ottawa facilities, thus reaffirming its view of the region's comparative advantage in research and development-oriented labour market.

Ottawa's Federal Agencies Towards Local Embeddedness

As noted above, two federal agencies in particular have contributed to Ottawa's particular model of development, although until recently neither of them had an explicit *regional* (or 'cluster'-oriented) economic development agenda. Firstly, the Communication Research Centre has had a significant presence in the region since 1969, when it was created out of the federal Defence Research Board as a civilian research centre. Currently employing more than 200 researchers in the western part of the Ottawa region, the agency is dedicated to applied research and development in communications and related technologies. Its special status encourages it to operate as an arms-length institution, thus enabling it to bridge its missions within the public sphere (primarily related to the support of defence-related research in telecommunications) and its activities as an important resource and catalyst for Canadian firms.

Despite the national outlook of this institution, it is clearly embedded in a regional context, not merely because of its location but importantly because of the involvement of several key local entrepreneurs in its programmes and policies. One example of such embeddedness was the appointment in the mid-1990s of Gerry Turcotte – one of Ottawa's key civic entrepreneurs, a leading advocate of a networking approach to innovation and technological development, and formally a founding president of the Ottawa-Carleton Research Institute – to lead the CRC and the activities of the CRC's Innovation Centre in support of small business start-ups and growth.

Secondly, the National Research Council has had a historically considerable impact on the creation and development of technology-oriented firms in the region. The contribution of the NCR, by its sheer presence in the region, continues to be a "major strategic asset in the region's capacity to encourage industrial innovation" (OED, 1997, p. 9). Its explicit mandate is to undertake, assist or support scientific and industrial research in different fields of national importance; to investigate standards and methods of measurements; and to work on the standardisation and certification of scientific and technical apparatus and instruments, and of materials used or usable by Canadian industries (NRC, 1997). During the 1990s, the activities of the NRC amounted to an impressive number of research projects, industrial support programmes, and innovation and networking initiatives, despite significant reduction of resources (to the tune of some 76.2 million Cdn$ in the early 1990s). Among these projects and activities one could note the NRC's emphasis on the transferring of technologies from its laboratories to industry for commercialisation via multi-partner initiatives.

During the 1990s, the NRC reported collaborations with over 375 firms, 60% of which were considered joint research projects (NRC, 1997).

Of particular importance has been the NRC's Industrial Research Assistance Programme (IRAP), which is designed to help Canadian firms to develop and exploit technologies. Based on a national network of Industrial Technology Advisors (ITA's), the IRAP demonstrates the increasing sensitivity to the role of clusters and spatial proximity in innovation systems and flexible, technology-intensive production complexes. For example, there are some 250 ITA's situated in more than 166 locations, and close to 90 cities and communities across Canada. It is estimated that the IRAP's contribution to firms amounted to 59 million Cdn$ (1995 value) for some 3,475 projects that were conducted on a cost-share basis, thus creating incentives for firms to invest in R&D, while externalising some of the risk associated with the innovation process. In addition, the emphasis of the IRAP has been on small and medium size enterprises that have fewer than 500 employees, based on the rationale that "these firms form the bulk of Canada's industrial base" (NRC, 1997, p. 10).

Indeed, towards the mid- and late-1990s, a somewhat refreshed approach to the geographic dimension and the importance of industrial growth and innovation capacity was apparent within the overall framework of the NRC's activities, perhaps representing something of a greater awareness toward the spatial embeddedness of firms, technologies and institutions:

> Innovation in Canada has been – and will continue to be – driven by regional and community-based systems. This means that our efforts to strengthen Canada's system of innovation must recognise the importance of geographic concentration, technology clusters, and more local initiatives (NRC, 1997, p. 1).

An important step toward enhancing the local embeddedness of federal institutions (the NRC in this instance), solidifying the federal role in an evolving *regional* mode of governance, and strengthening the associative economic framework of the Ottawa region, was taken in 1995. In December of that year, the NRC, the Ottawa Economic Development Corporation, and the Ottawa-Carleton Research Institute announced a strategy and plan to stimulate innovation and entrepreneurship in the National Capital Region, and strengthen the NRC's links to firms, financial institutions, industrial research agencies and other local interests. A key element in this strategy was the convening of a Technology and Innovation Roundtable Forum as a venue for Ottawa 'technology leaders' and other players to meet and discuss local barriers to entrepreneurship and technology development and "to begin a process of knocking them down" (Chhatbar and Darch, 1997, p. iv). In some sense, this might be interpreted as an active engagement of actors representing different spatial scales of governance (national and regional) in an effort to construct a distinctive regulatory milieu.

It might also represent a growing recognition by both local/regional actors and national agencies such as the NRC that industrial governance does require the *active* nurturing and fostering of 'relational assets,' 'untraded inter-dependencies' (Storper, 1997) and the 'soft institutional infrastructure' (Cooke and Morgan, 1998) which could contribute to the overall performance of a regional system of innovation. It is worth noting, for example, that the activities of the Action Plan for Innovation (coming out of the NRC/OED networking initiative) covered an impressive number of areas, including labour market governance and training initiatives, the most celebrated of which has been the *O-Vitesse Programme*, a pilot project to 'fast-track' training of software engineers, involving Mitel, the NRC, Carleton University, and the University of Ottawa; the encouragement of stronger links between firms; technology forecast roundtables; development of a 'virtual' research park; regular meeting between local firms and NRC agencies; the co-ordination of international business visitations in the Ottawa region; a more pro-active approach to the mobilisation of venture capital; and new initiatives to commercialise NRC technologies under the new NRC Entrepreneurship Programme.

Within the evolving institutional landscape related to the Ottawa technology-oriented complex, it is possible to identify a number of important political, normative, and even cultural underpinnings. Importantly, these political foundations have originated from a complex intersection of local entrepreneurs, organisational restructuring, and political norms. In addition, corporate cultural stances related to the nature of contemporary competition, technology and entrepreneurial drive have also featured rather prominently in narratives and actions related to institutional change, as exemplified by the growing involvement of the NRC in regional affairs and the social construction of a regional system of innovation.

However, while ideas regarding the importance of networking and collaborative institutional environment have clearly 'infiltrated' the action framework of individuals and agencies, the context of local institutional fragmentation and incoherence throughout the National Capital Region in general, and its flagship high-technology complex in particular, was an issue of considerable political concern at the time, and was referred to rather inaudibly in the Plan's discourse:

An innovative economy demands systematic approaches and strategies that efficiently mobilise and link powerful combinations of people, capital, resources and ideas. These resources, links and the resulting entrepreneurial activity constitute a 'System of Innovation...However, [*sic*] at present, innovative efforts and support for innovation are divided among a number of industrial, university and government institutions in the region. The Ottawa Economic Development Corporation and NRC have joined forces to begin the process of bringing these elements together through an action plan for accelerating the building of a coherent and powerful innovation system in the region (OED/NRC, 1995, p. 5).

Having said that, it should be recognised that the new institutions of the Ottawa innovative milieu, as exemplified in the creation of such bodies as the Ottawa-Carleton Innovation Forum, aimed at "accelerating the building of a coherent and powerful innovation system in the region" (OED/NRC Task Force, 1995, p. 5), were given important discursive support from a variety of sources. These discourses themselves represent significant insights regarding the process of institutionalisation – what a regional innovation system actually means in terms of institutional responsibilities, and the role of the state and civil society; and what interests fuel the process of institutional development.

OED and NRC Task Force members recognise that industrial innovation is largely a local phenomenon, frequently centred on clusters of firms...While local innovation systems are emerging across Canada, the National Capital Region has the potential to lead through examples of firms, universities, government laboratories and programmes, banks and financial institutions working together in networks and collaborative activities and through new approaches to encouraging those links (OED/NRC Task Force, 1995, p. 7).

Such narratives demonstrate the importance attributed to 'networking' and 'collaboration' as key concepts around which to build institutional practices of economic governance that would support the NCR as a regional innovation system.

The process of institutional change triggered by the NRC/OED initiative did, in important ways, signify the unwillingness of actors to await (technological and other) changes passively. Rather, as a member of the Innovation Forum remarked, regional pro-activism, induced by key civic entrepreneurs and expressed through institutional change, has reflected a desire to anticipate and even to influence change, to the extent that this is possible:

Innovation is the art of creating something new or the act of changing that which exists into something different. We live in a world of change...The only question you need to consider seriously is whether you want to be dragged along by change or whether you wish to drive change; do you want

> to ride the wave or be caught in the undertow? (Senior executive, high-technology firm).

And as other senior high-technology personalities argued when addressing the Regional Innovation Roundtable, the notion of 'local embeddedness' seems to have taken some very material form in the Ottawa region:

> With technology advancements being generated so quickly, no one, not even Nortel, can keep track of all the trends single-handedly...Other cities [*sic*] have the same aspirations of high-tech supremacy as we do. If we're to give the National Capital Region its best chance for future success in this industry, we'll have to forge stronger links among ourselves (President of Nortel Technology, addressing the Regional Innovation Forum in April, 1997).

> If we look at some of the best [high-tech regions], like the Silicon Valley in California, Research Triangle in North Carolina, and the Boston Area, perhaps we can get some clue. Silicon Valley was probably the most spontaneous, with the Research Triangle and Boston being driven somewhat more by incentives. In each case, there is located nearby one or more world class universities...For the Ottawa-Carleton region it is perhaps even more important to provide a pool of well-educated technical people to fuel an almost insatiable demand from our high-tech companies...Certainly, the NRC has been great help to [Mitel] in the area of processes related to the manufacture of semiconductors, and I applaud the move of the NRC to be even more responsive to the private sector than it had been in the past (Mitel Corporation executive).

However, without taking the comments quoted above lightly, it is important to identify the normative and political underpinnings that surround the very process of institutional change. These underpinnings have important implications on the representation of interests in and through associative institutions, and on the qualitative character of institutional performance. Thus, for example, one could note that despite the discourses of collective action and associationalism espoused by many within the framework of the Regional Innovation Forum, the overall attitude has remained suspicious towards the role of the state as an active agent within the overall framework of economic governance. Consider, for example, the following (and somewhat contradictory) remarks by a federal government official, who had participated in the Forum's deliberations:

> It is a pleasure to be here in that Government has a key role to play in the economy generally, in high-technology in particular, and in the local high-tech economy most particularly of all. But it is a different role than the one we are accustomed to. The role of Government is evolving from a doer to a facilitator...The type of government Canada needs is leaner, more effective,

> more flexible and more cost effective government. This new role requires co-operation and support between the Government and local communities (Marcel Massé, President of the Treasury Board, addressing the Regional Innovation Forum, April 1997).

Even more pronounced was the reaction to the overall apparatus of the Innovation Forum by one of the more important high-technology associations, the Canadian Advanced Technology Association (CATA). Although CATA would have been a natural, and indeed a central institutional partner within the Forum's activities and the other strategies formulated by the NRC and OED, it has chosen to take a very limited role in this experiment. The primary reason, as conveyed by a senior CATA official during an interview was the supposed 'heavy hand' of government which was associated with the Innovation Forum, having originated from national and regional state agencies (NRC and OED). Such 'heavy' government presence – despite the apparent discursive gestures to 'government as a facilitator' made by state officials – was deemed unacceptable to an institution that has espoused anti-government attitudes throughout its existence:

> We did not take a serious part in the NRC initiative, and I can't describe our relationship with OED as very close in any meaningful way…the reason for all of this is they all represent government…The Regional Innovation Forum is government-led, and we're not interested in government involvement. In my mind, any forum which has government in it is useless (senior official, CATA).

Thus, despite the apparent (and at least partially justified) excitement surrounding the establishment of the Regional Innovation Forum and other federal-regional frameworks for collaboration, and in spite of their contribution to greater dialogue among a variety of institutions, their potential to emerge as sustained associative governance mechanisms might be limited. In the first instance, as discussed above, cracks began to appear in the collaborative framework for dialogue when groups such as CATA and others within the technology-oriented cluster voiced their opinion that government should remain, at best, a minor partner, thus undermining NRC and OED's leadership in the forum. Secondly, the opposition to state involvement is contradictory because of the central role that government (especially federal research laboratories and agencies) has played in the development of the Ottawa high-technology cluster, and the continued importance of public institutions to the viability of the Ottawa economy in such areas as education and training, and infrastructure.

Expanding the Institutional Network
The Re-emergence of High-Technology Actors

By the mid-1980s a number of structural factors (economic recession, the growth of new technologies, political decentralisation and so on) coincided with local specificities (the presence of particular personalities and civic entrepreneurs from both the public and private sectors, for example) to produce a process of institutional change that would make an important imprint on the region's system of industrial governance. In its most pronounced form, this process led to the development of the Ottawa-Carleton Research Institute (now called the Ottawa Centre for Research and Innovation or OCRI) as a focal site from which to generate a 'new' regional identity. Since then, OCRI has been instrumental in projecting the region's success in establishing a thriving technology-oriented complex, and providing a forum for greater networking between firms, educational institutions and other organisations.

The drive towards the formal establishment of this institution began in 1983, when a few board members of the Ottawa Economic Development Corporation, a group of university officials from Carleton University, the University of Ottawa and Algonquin College teamed-up with a number of individuals from Ottawa's high-technology community, and decided to establish a high-technology research institute in the region, focusing primarily on the fields of micro-electronics, communications and computers (interviews with former OCRI officials). From its inception, OCRI rapidly became one of the region's key economic development agencies, having established itself as an important medium for public-private partnership oriented toward the technology sector. OCRI's networks of activities have reflected a growing perceptive need for local pro-activity in order to solidify Ottawa's place in an emerging so-called 'new' spatial division of labour, in which the region sought to establish itself as a 'recognised' technology complex alongside the likes of Silicon Valley and the Boston district.

The notion of a pro-active agency, combining resources from the public sector and industry to tackle a potential weakness in the governance of a regional labour market (the lack of 'reflexivity,' as defined by particular business interests, in university programmes), is of particular interest. This is because the shortage of adequate skills for particular industrial sectors would be socially constructed by discourses and practices as the key obstacle in the region's quest for world-wide status as a technology district. As such, the realignment of public and private institutions can be seen, at least partially, as a local regulatory experiment with respect to the governance of labour markets. As the Dean of Carleton University's commented at the time,

> As we see it our problem is one for the continued health of the whole Ottawa area high-tech operation – what's going to been needed is a lot more industry-university co-operation. [The institute] will give us in one spot the mechanism for the synergy that we need (CIDC, 1983, p. 17).

An important role in institutional change and the emergence of new actors was played by 'story telling': narratives and discourses emphasising the resilience, entrepreneurial drive and the collaborative nature of the advanced technology sector in Ottawa, and its role in re-defining regional identity.

> The past twelve months have presented new challenges to the region's high-technology firms. Small and large firms alike have sailed through rough waters and are now embarking on new ventures and new expansions which will enhance the region's reputation as a world class centre in the coming years (CIDC, 1983, p. 12).

And as the president and general manager of CIDC at the time proclaimed:

> Camaraderie among the high-tech industry is a unique feature of Ottawa-Carleton. Importantly, this camaraderie encompasses a community which understands and supports high-technology…By working together, the firms here overcome problems and nurture new future success stories (Marshall, 1983, p. 2).

The growth of OCRI as institutional foci for actors from the Ottawa technology-oriented complex and several key personalities from the region's educational institutions was rather impressive and swift. It was also an example of the region's ability to extend its network of existing institutions, since OCRI was, in a sense, "born in CIDC's offices" (interview with former CIDC official). In time, OCRI developed its own identity, both in spatial terms (identified with the region's west-end) and in its substantive-representative terms (echoing the interests of the technology-oriented fraction within the region's business class), and this in turn would lead to some considerable inter-institutional frictions which are subsequently explored.

By 1993, OCRI has completed its ten years of operation. Throughout those ten years, OCRI has evolved from a two-person operation, largely funded by the regional government, to 'a significant force within the region' with 14 staff members (RMOC Report Annual, 1993), and mostly financed by its members. In addition, examples of OCRI's success in 'thickening' the institutional network of the Ottawa-Carleton region are numerous, and include the following:

- *OCRInet*, a networking project to connect the major 'research nodes' of the region into an integrated fibre optic network, which was established in early 1993, and 'catapulted the region into the reality of being Telecommunications Valley North' (OCRI Annual Report, 1993). The development of this project took 3 years and significant volunteer work given OCRI's limited financial resources.

- *OPCOME/OCRI.* This project saw the establishment of a consortium of the Optical Processing and Computing Consortium of Canada with OCRI. Involving officials and industrial representatives from the local and the federal levels, the consortium's aim was to provide for a dialogue and collective action between the private and the public sector regarding innovation and commercialisation of technology in optics-related industry in the Ottawa region.
- *OCRI Technology Transfer Centre* (*TTC*). In 1993, OCRI successfully completed its lobbying efforts to assigning a senior technology transfer officer to its operation. The main goal of the Centre is to nurture research technology from its 'birth' in a university or research institute laboratory into a successful business. The development of the TTC could be interpreted as part of Ottawa-Carleton's emerging innovation milieu. It initially consisted of ten institutions dedicated to the advancement f technology opportunities in Ottawa-Carleton in the fields of science and engineering, medicine and pharmacy, and agriculture. Recognising the significant amount of research done in the region (estimated at 2 billion Cdn$ a year) by federal laboratories, universities and companies with large R&D budgets, the TTC has sought to transfer technology to either newly-set companies, or by licensing technology to existing firms. This institution, in itself a 'spin-off' of OCRI, also demonstrates the importance of localised social relations in the innovation process, and the institutional-infrastructural underpinning of a specific spatial pattern of technological development. As a TTC executive argued, "we're committed to increase, enhance and improve communication and the 'hit rate' of opportunities in a regional context. Our goal is to ensure that as many technologies are being exploited as regionally as possible."
- *The OCRI Research Management Board* (*RMB*). The RMB has been part of OCRI since its inception. It acts as reaffirming research direction, acting as a sounding board and an advisory group. In time, the RMB has evolved into a body that acts as a vehicle for the provision of information on OCRI activities to its members. Its board of directors does signify something of a wide-

ranged 'technologically oriented spatial coalition,' with representation from Ottawa's three major educational institutions, representatives from financial institutions, the CRC, leading information technology firms and Bell Northern Research. The specific role of the RMB is to act as an advisory council to OCRI on strengthening ties with the local research community covered by the OCRI mandate; develop initiatives of direct interest to members; and identify opportunities for joint research among the members.

Thus, throughout the 1990s OCRI has established itself as an important institution in the region via its networking strategy and various networking 'events' which have facilitated discussions and forged community identity. In addition, OCRI's activities could be interpreted as attempting to insert itself as an institution which bridges markets and public institutions in one particular area: the governance of the Ottawa labour market in an attempt to redress the concern of high-technology firms regarding skills shortages. Within the framework of this particular venture, the inclusion of some prominent personalities from the universities and colleges in the activities of OCRI and other technology-oriented associations (such as the Ottawa Life Sciences Council or OLSC) was indeed judged by many players to have contributed to the region's emerging regulatory environment.

Furthermore, there has been a growing sense that institutions such as Carleton University and the University of Ottawa have indeed responded to the cry-out of technology-oriented actors regarding the problem of skill shortages in the region, and restructured their academic programmes in a way that would be more in tune with the demands of firms. Interestingly, this growing sense has reflected the success that initiatives such as the Innovation Roundtable have had in putting the technology 'social bloc's' labour market governance agenda on the map, and has been confirmed by players from the private sector and academic institutions.

> In the last few years the private sector has been involved in establishing training initiatives that would correspond to the gaps identified in the regional labour markets…I have seen enormous growth in short courses related to information technologies, and places like the Toronto Schools of Business and The Information Training Institute, their volume of business sky-rocketed...The university level programme – *OVITES* – has been a highly successful model: the universities are happy, the students are happy, and the companies are happy. So certainly, at various levels moving from short courses to year-long courses, there has been reaction and adoption to the labour force requirements (Senior executive, high-technology firm, and a senior board member of OED).

Thus, economic governance institutions in the region have established themselves as important mediators between different segments of the industrial community. These intermediary institutions have had a considerable role to play in soothing the inherent tension between rogue individualism usually associated with a North American enterprise culture and the advantages of collective learning that has been increasingly recognised by actors as instrumental to the region's positioning within a wider political economy.

> At the heart of community building is a new style of civic leadership, which…uses the spirit of enterprise to restore the spirit of community. Civic entrepreneurs are characterised by their entrepreneurial traits as risk takers, and as people of conviction, vision and energy. They are also networkers, educators, facilitators, and catalysts for civic change. They interface between business, government, and society at large and they demonstrated a collaborative style that expresses a philosophy of inclusion and that empowers others to take action. They are a pivotal link between the economy and the community and are often led by technology entrepreneurs who have grasped the connection between collaboration and innovation (OCRI, 1998, p. 1).

Of course, there may be considerable exaggeration in these views on the extent and depth to which entrepreneurship and networking have co-existed in Ottawa. Nor is the issue of inclusive economic governance unproblematic. For instance, associative dynamics and genuine collaboration between labour unions, community groups and business representatives have remained rare. Whatever attempts at such collaboration existed, they have been undermined by deep mistrust and were short-lived (Leibovitz, 2001). In addition, the bridging of the Anglophone part of the National Capital Region with the Francophone community of Gatineau on the Québec side has remained, at best, tentative and under-developed (Leibovitz, 2001). The collaborative atmosphere induced by Ottawa-based institutions – mostly in the Ontario side – has been therefore selective and limited.

Yet, in fairness, as we have seen, the institutional development of Ottawa does exhibit some tendency towards patterns of associationalism. The evidence from the role of institutions in building an associative atmosphere throughout different segments of the economic community, as opposed to one industrial sector, and on a region-wide scale (spanning both the Québec and Ontario parts), as opposed to Kanata's 'west-end' technology cluster, is quite mixed (Leibovitz, 2001). While the historical and path-dependent nature of Ottawa's economic transformation has left an important legacy of associationalism within the industrial milieu of the technology-oriented cluster, associative dynamics in governance have not diffused substantially to different sectors and different communities throughout the region.

Conclusions

Institutionalist approaches have often contributed a perspective on urban and regional economic change that is well grounded in wider social, political and cultural contexts. Yet, they have often neglected the extent to which periods of economic transition are marked by interests, values and power. In short, they have been deficient in pointing to the role of the state in the mediation of policy choices. For the governance of economic change to be more fully appreciated, one needs to consider how selectivity as well as reflexivity are intertwined.

Canada's National Capital, Ottawa, has experienced dramatic economic change in the past 50 years, but it has been more accentuated in the past 15 years with the impressive growth of its technology-oriented complex. In this chapter, it has been argued that this rise has been closely interwoven with a remarkable conjuncture of 'high' level institutions (a complex array of federal government agencies) and local initiatives and actors. Together, the cajoling of interests and institutions has been instrumental in underpinning the construction of a social capital-inducing environment: the Ottawa high-technology complex has exhibited considerable traits of community identity, a sense of mission and enterprise, and collective action. Ottawa's distinctiveness has been sometimes taunted as signifying its irrelevance to the rest of Canada's communities. Yet, in this chapter, we have seen that a degree of institutional 'thickness' and collective action can be generated if and when the appropriate circumstances materialise. It is true that Ottawa enjoys an institutionally-thick environment, but the federal government has the capacity to trigger positive dynamics across other Canadian communities.

At the same time, there has been much selectivity in the vision emanating from this type of reflexivity. There has been less then impressive cross-'border' strategic governance: the Ontario and Québec parts of the National Capital Region have found it difficult to engage in a region-wide endeavour of collaborative governance. Within the region, the state of the older manufacturing base, internal inequalities (for instance between central city communities of Vanier and central Ottawa and the prospering suburbs), and the tension between public sector trade unions and high-technology actors have all pointed to the relatively narrow version of institutionally-based social capital.

During the early years of the 21st century, the high-technology sector suffered a humbling blow to its perceived resilience, with the burst of the so-called dot com bubble. In Ottawa, the number of employees in its high-tech establishments dropped sharply from some 82,000 to 69,000 (Valiquette, 2006). Beyond the obvious question of whether, even under the best of circumstances, regional institutions can ever hope to truly achieve a negotiated system of economic governance, the crisis has raised questions about the genuine commitments of both firms and supporting institutional actors to engage in reflexive governance around issues of labour training and collective business strategies (Valiquette, 2006). The limits of associative regional governance have thus been demonstrated, but the question remains whether past and current practices are adjusted to take a broader view of Ottawa's diverse communities.

References

Agnew, J. (2000). From the political economy of regions to regional political economy. *Progress in Human Geography*, *24*, 101-110.

Amin, A. (1999). An institutionalist perspective on regional economic development. *International Journal of Urban and Regional Research*, *23*, 365-378.

Amin, A. (2004). Regions unbound: towards a new politics of place. *Geografiska Annaler*, *86* (B), 33-44.

Amin, A., Thrift, N. (1995). Globalization, institutional 'thickness' and the local economy. In P. Healey *et al.* (eds.), *Managing Cities: The New Urban Context*. Chichester: Wiley, 91-108.

Arthur, W.B. (1994). *Increasing Returns and Path Dependence in the Economy*. Ann Arbor, MI: University of Michigan Press.

Chhatbar, A., Darch, M. (1997). Forward. In Chhatbar, A. and Darch, M. (eds.). *Inaugural Technology Forecast Roundtable Report*. Ottawa: National Research Council.

Commercial and Industrial Development Corporation of Ottawa. (1983). *Annual report*. Ottawa: CIDC.

Cooke, P. (1995). Keeping to the high road: learning, reflexivity and associative governance in regional economic development. In P. Cooke (ed.), *The Rise of the Rustbelt*. London: UCL Press, 231-245.

Cooke, P., Morgan, K. (1998). *The associational economy*. Oxford: Oxford University Press.

Cooke, P., Uranga, M.G., Etxebarria, G. (1998). Regional systems of innovation: an evolutionary perspective. *Environment and Planning A*, *30*, 1563-84.

Cumbers, A., MacKinnon, D., McMaster, R. (2003). Institutions, power and space: assessing the limits to institutionalism in economic geography. *European Urban and Regional Studies*, *10*, 325-342.

Doyle, D. (1996). *An Overview of Ottawa-Carleton's High Technology Industry*. Ottawa: Doyletech Corporation.

Gertler, M.S. (2001). Technology, culture and social learning: regional and national institutions of governance. In Gertler, M.S. and Wolfe (eds.). *Innovation and Social Learning*. Basingstoke: Macmillan.

Gertler, M.S., Wolfe, D. (1998). The dynamics of regional innovation in Ontario. In De La Moth, J. and Paquet, G. (eds.) *Local and Regional Systems of Innovation*. Amsterdam: Kluwer, 211-238.

Harrison, R.T., Cooper, S.Y., Mason, C.M. (2004). Entrepreneurial activity and the dynamics of technology-based clusters. *Urban Studies, 41*, 1045-1064..

Henry, N., Pinch, S. (2000). Spatialising knowledge: placing the knowledge community of Motor Sport Valley. *Geoforum, 31*, 191-208.

Hirst, P. (1994). *Associative democracy*. Cambridge: Polity Press.

Keating, M. (2001). Rethinking the region: culture, institutions and economic development in Catalonia and Galicia. *European Urban and Regional Studies, 8*, 217-234.

Langendijk, A. (1999). The emergence of knowledge-oriented forms of regional policy in Europe. *Tijdschrift voor Economische en Sociale Geografie, 90* (1), 110-116.

Lawson, C., Lorenz, E. (1999). Collective learning, tacit knowledge and regional innovative capacity. *Regional Studies, 33*, 305-317.

Leibovitz, J. (2001). Associative governance? Constructing institutional change in two Ontario city-regions. Unpublished PhD Dissertation, The University of Toronto.

Leitner, H., Sheppard, E. (2002). The city is dead, long live the net: harnessing European international networks for a neo-liberal agenda. *Antipode, 34*, 495-518.

Lovering, J. (1999). Theory led by policy: the inadequacies of the 'new regionalism.' *International Journal of Urban and Regional Studies, 23*, 379-395.

Lundvall, B-A., Maskell, P. (2000). Nation states and economic development – from national systems of production to national systems of knowledge creation and learning. In Clark, G.L., Feldman, M.P. and Gertler, M.S. (eds.). *The Oxford Handbook of Economic Geography*. Oxford: OUP, 45-56.

MacLeod, G. (2001). Beyond soft institutionalism: accumulation, regulation, and their geographical fixes. *Environment and Planning A, 33*, 1145-1167.

Mallett, J.C. (2002). *Silicon valley north: the formation of the Ottawa innovation cluster*. Ottawa: The Information Technology Association of Canada.

Marshall, W.A. (1984). 'Forward' in CIDC's *"Interface Ottawa '84"* (*Annual Report*).

Martin, R. (2000). Institutionalist approaches to economic geography. In Sheppard, E. and Barnes, T. (eds.). *Companion to Economic Geography*. Oxford: Blackwell, 77-97.

Maskell, P., Eskelinen, H., Hannibalson, I., Malmberg, A., Varne, E. (1998). *Competitiveness, localised learning and regional development*. London: Routledge.

National Research Council. (1997). *Report to Parliament 1995-1996.* Ottawa.

Ottawa-Carleton Economic Development Corporation and National Research Council (1995). *Action Plan for Innovation in the Ottawa-Carleton Region.*

Ottawa Planning Department. (2005). *Ottawa 20/20: economic strategy*. Retrieved on July 8th, 2006 from www.ottawa.ca/city_services.

Ottawa Economic Development Corporation. (1997). *Succeeding globally through collaboration: strategic economic development action plan, 1997-1998.*

Parr, J.B. (2005). Perspectives on the city-region. *Regional Studies*, *39*, 555-566.

Piore, M., Sabel, C. (1984). *The second industrial divide: possibilities for prosperity*. New York: Basic Books.

Putnam, R. (1993). *Making democracy work*. Princeton: Princeton University Press.

Regional Municipality of Ottawa-Carleton. (1996). *Economic trends and their implications for Ottawa-Carleton*. Ottawa: Planning and Property Department, Policy Division.

Saxenian, A-L. (1994). *Regional advantage: culture and competition in Silicon Valley and route 128*. Boston: Harvard University Press.

Scott, A.J. (1996). The collective order of flexible production agglomerations: lessons for local economic development policy and strategic choice. *Economic Geography*, *72* (3), 219-313.

Steed, G.P.F., DeGenova, D. (1983). Ottawa's technology oriented complex. *Canadian Geographer*, *XXVII*, 263-278.

Storper, M. (1995). The resurgence of regional economies, ten years later. *European Urban and Regional Studies*, *2* (3), 191-221.

Storper, M. (1997). *The regional world*. New York: Guilford Press.

Storper, M., Scott, A.J. (1989). The geographical foundations and social regulation of flexible production complexes. In J. Wolch and M. Dear (eds.). *The Power of Geography*. Boston: Unwin Hyman, 78-97.

Swygedouw, E. (2000). Authoritarian governance, power and the politics of rescaling. *Environment and Planning D: Society and Space*, *16*, 63-76.

Taylor, J.H. (1986). *Ottawa: an illustrated history*. Toronto: James Lorimer & Company.

Valiquette, L. (2006). Editors file: self-preservation is mother of invention, *The Ottawa Business Journal*. Retrieved on July 5th, 2006 from www.ottawa businessjournal.com.

Part IV

The Unique Ottawa Cluster
Regional, Bilingual, and Cosmopolitan

CHAPTER 10

A Cross-Border High-tech Cluster

A View From Gatineau[1]

Guy CHIASSON

Département des sciences sociales,
Université du Québec en Outaouais

Introduction

The Ottawa region's technology sector is considered by many as one of the most successful high-technology (HT) clusters in Canada. Shavinina (2004) in a recent book enthusiastically calls Ottawa "the high-tech capital of Canada" and "Silicon Valley North", suggesting that Ottawa is no longer only the political capital but also a driving force in the Canadian new economy. This apparent success has also triggered the interest of many researchers, as this book suggests, and is the topic of a burgeoning literature (Harrison, Cooper, Mason, 2004; Madill, Haines, Riding, 2004). While this literature has extensively studied different important aspects of the 'Ottawa cluster,' it has up until now not given much attention to what is happening on the Québec side of the Ottawa River, that is, in the City of Gatineau[2]. Some studies seem to have altogether forgotten Gatineau's high-technology (HT) presence and concentrate solely on the Ontario side of the river. Others talk about "the Ottawa region," the "National Capital Region"[3] or the "Ottawa

[1] This paper counted on the careful assistance of Kymberly Allard-Skerkowski.

[2] The new City of Gatineau was created by the merger in 2002 of 5 former municipalities: Aylmer, Gatineau, Hull, Masson-Angers and Buckingham. This amalgamation represents the urbanized core of the Outaouais region.

[3] Since at least the 1950s, the National Capital Region (NCR) has included the Québec side of the Ottawa River. In terms of science and technology, Statistics Canada treats the National Capital Region as one statistical entity, which means that a lot of available data is at that level (Godin, 2000).

cluster" while seeming[4] to lump Gatineau and Ottawa together. In both cases the results are similar: we know little about the specifics of the Gatineau HT presence nor about the cross-border relations within the Ottawa cluster.

This oversight is, of course, quite understandable. Gatineau's HT sector is in many ways a dwarf when compared to the Ottawa one. With about 40 companies employing some 3,000 people in 2001 (FM Consult, 2002), the Québec portion is a minor player compared to Ottawa's much larger industrial base. Regardless of its status, Gatineau's new economy sector is not negligible. Some consider it the third most important technology cluster in the province of Québec (Doucet, Favreau, Robitaille, 2005) and it is without a doubt an important lever in the local development strategy put forward by municipal (for instance the *Corporation de développement économique de Gatineau*, an arm's length municipal economic development corporation) and regional actors (the former *Conseil régional de développement de l'Outaouais* as well as the *Société de diversification économique de l'Outaouais*) for the Québec side of the river.

The aim of this chapter is to address Gatineau's HT sector while reflecting on the cross-border dynamics within the 'Ottawa cluster.' Our research assembled and analysed public documents from a wide sample of institutions involved in Gatineau's HT (see *Appendix A* for a list of these institutions). This research builds upon data from what is possibly (regarding existing knowledge) the only two recent research exercises conducted specifically on Gatineau – one led by a Québec government agency (Conseil de la recherche et de la technologie du Québec, 2001) and another by a local technology development agency (Materazzi, 2004) – in order to get a better grasp of its emerging new economy. First, we take a closer look at some current studies of the Ottawa new economy complex in order to show how they have forgotten to take into account the significance of the Québec-Ontario border. Our presentation of Gatineau's HT sector is divided into two parts: one concerned more with the dynamics at the level of local businesses and another devoted to presenting governance by public and semi-public institutions. In both instances (local businesses and governance institutions), our intent is to

4 In some cases it is not totally clear if the "Ottawa cluster" or the "Ottawa region" actually includes the Québec side or not and if the studies are conducted on both sides of the border or only on the more visible Ontario-side HT sector. Even if the study is set in the National Capital Region, authors will usually focus on companies, institutions and partnerships (OCRI for instance) that operate on the Ontario-side leaving Gatineau's HT unexamined.

highlight, from the perspective of Gatineau, the cross-border dynamics in terms of integration as well as in terms of separation and conflict.

Of High-tech Clusters and Institutional Boundaries

Many contemporary regional development scholars, geographers and other social scientists have tried to understand the spatial dynamics underlying the new economy. For many of these authors, HT is not only an important part of the rising *post-Fordist* economy (Andrew, Houle et Thériault, 1992) but also suggests some new relationships between economy and space. For this very rich literature, in the old Fordist economy, firms were seen as fairly independant from territorial dynamics[5] (May, 1986; Scott and Storper, 1992), while in the new economy and more so the high-tech sector they are considered as much more grounded in the particular characteristics of local spaces (Storper, 1997; Maillat, 1992). Knowledge, information flows, human and social capital, and entrepreneurial culture are only a few of the many ingredients that may arise (or not) from local territorial dynamics and that are crucial to ensure the capacity of local private firms to produce the innovative products and services of the new economy.

The terms used to describe these new local (or regional) spaces tend to vary. In the 1980s, Philippe Aydalot (1986) and other European regional scientists (Proulx, 1992; Maillat, 1992) launched the concept of *milieux innovateurs* (innovative milieux) while studies on Italy revived Alfred Marshall's idea of an *industrial district* (Piore and Sabel, 1984; Tremblay, 1998; Becattini, 1992; Pecqueur and Rousier, 1992). These authors and others have also frequently used the terms *territories* in order to imply the new dynamics between local spaces and the economy. In Canada, David Wolfe (see Wolfe and Holbrook, 2000) has launched a major cross-Canada study on *regional innovation systems*. Finally, Micheal Porter (1990) uses the term *cluster* to describe the interconnections between different firms operating in spatial proximity[6].

---

5 For some, this explains why Fordist type firms can (fairly) easily relocate in order to look for cheaper labour or more flexible environmental and social laws or at least use this possibility to relocate in order to negotiate with local authorities. Alain Lipietz and Danièlle Leborgne (1992) call this strategy *defensive flexibility* (see also Tremblay, 1995 and 1998).

6 The term *cluster* is, however, not without its share of ambiguities (Harrison, Cooper and Mason, 2004). If it is often used to determine a spatial proximity such as the Ottawa cluster it can also be used to describe a particular sector in the whole high tech business: the 'life sciences cluster' or the 'information technology cluster' for example (ICF Consulting, 2000). This paper will prefer the geographical sense.

The scales referred to by these terms might also vary from very large regions (provinces such as Québec and Ontario as in the case of *regional innovation systems*) to much more local boundaries. However, they all pinpoint to the embededness of the new economy in sub-national spaces. We prefer to use the term *cluster* simply because it seems to be the preferred terminology among students of high-technology in Ottawa as well as many actors in the field (Harrison, Cooper, Mason, 2004).

Of course, the local spaces or clusters of the new economy are *functional* spaces (Fontan, Klein and Tremblay, 1999). That is, they do not *always* and most likely will not coincide with institutional boundaries (be they municipal, regional and potentially national). The often-studied and imitated Silicon Valley technology cluster (Shavinina, 2004; Saxenian, 1994) does not fit neatly into any local jurisdiction. What makes it a *cluster* is not a shared political or institutional space but more likely the density of interconnections or networks (between firms as well as between firms and public or civil society actors) that are allowed by geographical proximity. In the case of Silicon Valley as in other clusters, these networks can lead to a very visible form of collaboration: the sharing of some valuable collective resources (research centres, universities, hyperqualified workforce etc.). The recognised importance of networks have also brought some researchers to study some less tangible but nonetheless crucial aspects of clusters: a dynamic entrepreneurial 'atmosphere' as pointed out by Saxenian (1994) or a culture of 'excellence' as indicated by Shavinina (2004) among others.

This type of perspective focused on clusters (or however the authors might like to call them) has been present in studies on the Ottawa region high-tech economy. It has led Harrison, Cooper and Mason (2004) to study the cluster's different mechanisms to promote new businesses (attracting entrepreneurs through *magnet organisations*, incubating start-ups etc.), and Paquet, Roy and Wilson (2004b) to analyse the collaborative strategies put forward by public and private actors alike in Ottawa (-Gatineau?) to promote a highly-skilled workforce. Madill, Haines and Riding (2004) explore the local technical and social networks as a source of technical knowledge, business advice and venture capital while Chaiton, Dibbits, Paquet, Roy and Wilson (2000) examine the workings of regional governance, more precisely the role of three 'civic entrepreneur organisations' (the Joint Venture Silicon Network, the Glasgow Education-Business Partnership and the Ottawa Centre for Research and Innovation) in order to determine their ability to promote collective learning and innovation capacities. In short, recent studies have tended to focus on networks and collaborative governance relationships between public, private and tertiary sector actors. These spatially grounded networks are shown to be crucial in many ways to the cluster's performance.

This approach in terms of cluster has definite merits. For one, it serves as a reminder that the conditions driving economic growth and innovation in a post-Fordist economy are as much localised as they are triggered by national public policies (Wolfe, 2003). By insisting on portraying the Ottawa high-tech sector as a coherent whole – or in other words as a *functional* space – most authors have downplayed one aspect very specific to the Ottawa region cluster: the fact that it sits on the border between two provincial jurisdictions. As we suggested earlier, that border seems to disappear behind the cluster perspective. Studies have either implicitly considered the provincial border to be irrelevant to the dynamic of the high-tech economy or as is the case of Paquet, Roy and Wilson (2004), the lack of cooperation across that border is lamented. The theme of the border is present in a certain number of studies, but more often than not it refers to the traditional borders between the public, the private and community actors. When, for instance, Chaiton, Dibbits, Paquet, Roy and Wilson (2000, p. 5) talk about 'boundary-crossing relationships,' they refer to partnerships across these sectors and do not seem concerned by the issue of the provincial border:

> Since, all sectors are increasingly engaged in a variety of multi-stakeholder forums, there exists a growing inter-dependence across public, private, civic and educational spaces. No single organization seems to go it alone any more, and consequently both the manner by which cross-sector interactions occur, and the nature of the collective intelligence that emerges from them, become critical determinants of the quality of collaborative action and local innovativeness.

My intent here is not in any way to deny the existence of a cluster that spreads across the Ontario-Québec border and the importance of strong cross-border linkages in the high-tech industry. My concern is rather to reflect on the complicated cross-border relationships within the cluster. The fact that the Ottawa cluster is divided by a provincial border, on top of being the national capital, most likely complicates the situation by adding a lot of institutional players (municipal, provincial and federal) who have a definite interest in the high-tech sector's interactions across that border. How do these institutional players and the local civil societies connect and compete across that border?

These questions are essential in order to better understand Gatineau's place in the local technological scene. They are also interesting from the point of view of contemporary social and political theory. At least since the pioneer work of Manuel Castells (1996) on the *Network*

Society and Bertrand Badie's *La fin des territoires* (1995), social scientists have been concerned with the impact of networks on political and institutional spaces. In Castells' world, networks create *spaces of flows* that tend to cut across national institutional territories therefore diminishing the former prominence of States and their borders. Of course, information technology is a crucial component in the building of these new spaces of flows. Joachim Blatters' (2004, p. 530) comment on Castells' work confirm this point of view:

> Manuel Castells is not the only observer who sees revolutionary transformations taking place at the end of the twentieth century and the beginning of the twenty-first century. Driven by the dynamics of information technology, he envisions the 'rise of the network society,' where a 'space of flows' is superseding the traditional 'spaces of place' as the dominant logic for social organizations and institutions.

The relevant question for many contemporary researchers is not if States and institutional borders disappear behind the actions of networks as many have assumed in the past (Badie, 1995). Regardless of the drastic changes in their mandate, local institutions as well as provincial borders are not on the verge of disappearing (Biarez, 1997). The more likely question is how do these institutions adapt and interact with the new networks dynamics and the new *functional and cooperative* spaces portrayed by Castells and others (Fontan Klein and Tremblay, 1999)? Of course, as some have argued (Blatter, 2004), cross-border regions and cities such as the Ottawa-Gatineau region are prime spots[7] to observe the complicated interplay between *functional* and *institutional* territories. Our aim here is to expose some of that complexity through an analysis of Gatineau's (that is the institutional territory) place in the Ottawa cluster.

Gatineau as an Outpost of the Ottawa Cluster Integration and Dependency

The title of this section suggests that the Gatineau HT sector is very dependent on Ottawa's dynamism. That is quite correct if one only looks at the level of the firms and the private sector. Up until now, firms which operate in Gatineau are in many ways very dependent on their proximity to Ottawa. Our analysis addresses three dimensions that facilitate integration and sometimes perpetuate dependence of Québec's

---

[7] Blatter calls these border regions 'laboratories' were the dynamics of glocalisation play out.

firms in the cluster: the relative lack of research facilities, the production sub-sectors and firm mobility.

The rise of a significant technological sector in Gatineau can be of some surprise when one considers its economic history. Gatineau's economy was until the 1970s dominated by the wood industry (Gaffield, 1996). In this decade, the dramatic rise of federal government activities replaced the crumbling forest industry as the driving force of the economy allowing the region to become a leading service economy in Québec and definitely propulsing Gatineau into the postindustrial era (Doucet, Favreau and Robitaille, 2005).

In the case of Ottawa (Tremblay, unpublished), the federal government has played, since the 1950s, a key role in the birth and construction of the HT sector. The federal government acted not only as a major consumer of technological products, but also as a major provider and funder of Research and Development (R&D) facilities and capacities (Madill, Haines and Riding, 2004). In the case of Gatineau, the nascent HT complex was much more modest and could not, at least until the 1990s, challenge Gatineau's profound dependence on public service employment. Somewhat ironically, it is only when public sector employment slowed down in the early 1990s that many regional actors (that is from the Outaouais region) came to see the technological sector as a prime avenue for regional economic diversification (Doucet, Favreau and Robitaille, 2005; Société de diversification économique de l'Outaouais, undated).

Research Facilities

The *Conseil de la science et de la technologie du Québec's* (a Québec government-sponsored agency concerned with technological research and promotion) study concluded in 2001 that the Outaouais region had a fairly low potential for technological advancement compared to other regions in Québec. It had in its territory few research facilities (either public or private) while its university (the *Université du Québec en Outaouais*) produced very little technology-related research (Conseil de la science et de la technologie du Québec, 2001). Regardless of these shortcomings, the region had one of the strongest showings among Québec regions in terms of technological firms as well as residents employed in the HT (FM Consult, 2002). This apparent paradox shows the extent of the dependency of Gatineau's firms on Ottawa's research capacity.

Along the same lines, FM Consult's survey responded to by 29 of the 40 firms on the Québec side indicates that these firm have links with research centres located outside of the Outaouais region. While some of them (6) had links with universities outside of the region and others (3) had partnerships with federal government labs, none indicated that they maintained a partnership with a lab from Gatineau (F.M. Consult, 2002, p. 44). Added to this, the biggest high-tech employer in Gatineau (Bell-Noxia) does not have any research activities at its Gatineau site. These elements indicate that in terms of research, the cross-border exchanges are mainly unidirectional: firms from Gatineau rely mostly on Ottawa's research potential. This is not to say that there are no R&D activities and facilities[8] in Gatineau; rather, they have not been able to offset local firms' strong reliance on cross-border expertise.

Production Sub-sectors

The close integration (and dependence) between Gatineau and Ottawa firms can also be measured by the production sectors that dominate on both sides of the river. As Materazzi (2004) points out, the 40 companies from Gatineau[9] are all involved in information technology (IT). According to Materazzi's (2004, p. 244) classification, there were (in 2003):

- 8 firms in Telecommunications;
- 5 in Databases and Geomatics;
- 5 in Equipment and Microelectronics;
- 6 in Management Software;
- 11 in Internet and Multi-media;
- 5 in Computer Services; and,
- 3 in Security-related software.

IT has always been a strength in Ottawa firms. As Tremblay (undated, p. 6) points out, the:

> greatest single influence on the development of the Ottawa cluster, however, was the decision by Northern Electric (Nortel) to concentrate its R&D activity in the nation's capital via two subsidiaries – Bell Northern Research (BNR) and Microsystems International. BNR's labs and test facilities seeded the development of Ottawa's telecom expertise…

Even within IT, the sub-sectors where Gatineau's firms are present are the same as in Ottawa.

According to a report published by ICT Consulting (2000) and often cited in the literature, Ottawa HT business has strength in the following

---

8 Our last section will present some of the more recent R&D initiatives from Gatineau.

9 These 40 companies are the ones that have more than 5 employees.

sub-sectors: Telecommunications Equipment, Microelectronics, Software and Communications, Photonics and Professional Services. These are very much the same ones identified in Gatineau. Materazzi (2004, p. 244) sums it up nicely when he states that:

> technology companies in Gatineau are active in the same areas of information technology as Ottawa companies, but […] some sectors such as microelectronics are less well represented than in Ottawa. For instance, the absence of the life sciences sector is also evident. Furthermore, no sector has a significant critical mass. Thus, for example, companies that have been listed together under the heading of Internet and multi-media offer very different applications, including training, e-commerce, Website development, Internet service and so on.

From this comment, it becomes fairly clear that the Gatineau complex is not really self-sustaining since not only do companies have to rely on Ottawa's research capacity, but also because these firms don't seem to have been able to consolidate into critical masses in their respective sectors.

Firm and Employee Mobility

The fact that most companies from Gatineau share Ottawa's production sectors is quite understandable when one considers the strong cross-border mobility of firms and employees alike. Labour mobility across the border has been a fact of life in Ottawa-Gatineau for a long time (Doucet, Favreau et Robitaille, 2005). At least since the 1970s, public servants from either side of the border have crossed the river daily *en masse* to go to work in government (or private) offices.[10] This trend is undeniably present in the high-tech sector as well. More than half (52%) of people employed in Gatineau firms reside on the Ontario side while many Québec residents work in the Ontario firms or public labs. This means that Gatineau and Ottawa businesses share a common specialised workforce. For Gatineau's firms, it also means that they have access to much more important specialised labour and that they can benefit from

---

[10] Paquet, Roy and Wilson (2004, p. 231) report that the "intermingling of work and living patterns results in thousands of people crossing the river-border each day. According to a 1995 study by the provincial government of Québec, on any given weekday nearly 40,000 people left their home on the Gatineau side of the river to go work in Ottawa during peak commuting hours: conversely, over 14,000 people travelled from Ottawa to work on the Gatineau side of the river, figures that remain relevant benchmarks today, given the rise and subsequent contraction of technology-based industries in the late 1990s."

some of Ottawa initiatives in terms of workforce development.[11] It also means that they have to compete with a lot of the much bigger firms operating on the Ontario side. Interestingly enough, language does not seem to amount to a major barrier in labour mobility, since 48% of HT employees in Gatineau are anglophones (FM Consult, 2002, p. 41).

Firm cross-border mobility is also an important aspect of Gatineau's HT sector. Over the years, the Québec side has lost a certain number of companies which decided to close shop and move to Ottawa. The most famous and significant case is the moving of Alcatel which led to important job losses for the Gatineau HT sector (FM Consult). The jobs lost with Alcatel along with other closures on the Québec side were replaced over the years by new companies: a few of them are bigger firms locating some of their operations in Gatineau and many others which are new companies. "[M]ore than 75% of the technology enterprises in Gatineau were started by Ottawa-based entrepreneurs or by Gatineau residents who had previously worked in Ottawa companies or institutions" (Materazzi, 2004, p. 242).

Harrison, Cooper and Mason's (2004) argue in their study of the Ottawa cluster that in many cases, entrepreneurs come from outside the cluster and not only from other local firms. In Gatineau's case, this seems to be less the case. The figures presented suggest that Gatineau entrepreneurs mostly come from the cluster where they probably have gathered valuable technical knowledge, contacts and experiences before creating their own "startups" in Gatineau. The well know phenomenon of former employees creating their own firms in the same cluster has a different spin in Gatineau where it takes on a cross border dynamic.

Institutional Practices. A Tale of Two Cities[12]

The previous section has argued that at the company-level, Gatineau's new economy is in many ways integrated and dependent on Ottawa. It is in that sense that we called Gatineau an 'outpost' of the Ottawa cluster. Does this strong integration mean that the provincial border is irrelevant to the HT cluster dynamics?

---

[11] See for instance Paquet, Roy and Wilson's (2004) paper on *TalentWorks* an initiative aimed at workforce development.

[12] This section presents the first results of a study conducted on the respective roles of local, provincial and federal institutions on local HT policy. Our study includes an analysis of public documents (websites and public documents) produced by these institutions as well as some forthcoming interviews. This paper presents the initial results from an analysis of documents from those institutions involved in Gatineau.

While at the firm-level, integration is very visible and it might not be so much the case if one embraces the *governance*[13] of the high-tech sector. Focusing on governance implies taking into account at least the role of public institutions as well as private firms involved in the sector. The key role of public actors in local technological development has been recognised by many (Wolfe and Holbrook, 2000; Andrew, Houle and Thériault, 1992). Our concern here is to see how public institutions operating on the Québec side interact within the cluster and the significance of the border in their actions.

Provincial and Federal Actors

Many institutions operating at different scales of government are involved in the governance of Gatineau's new economy. The most well known contribution comes from the Québec government in the form of a tax credit on research and development activities. Gatineau's firms (like other firms from Québec) can have access to a 40% credit which they accumulate to the 35% credit already available through federal government taxation. The *Association de la recherche industrielle du Québec* (Québec association for industrial research), a network of innovative businesses operating in Québec, has opened a bureau in every Québec region in order to facilitate giving local firms access to the Québec tax credit (ADRIQ, 2005). Some authors point out rightly that this fairly generous credit has not been able to offset Ottawa's quasi-monopoly and the spatial concentration of growth in Ottawa's west end.[14] However, for others, it has still played a significant role in firm development (Materazzi, 2004). For sure, this credit is a major booster in Gatineau's HT promotion strategy.

A variety of other public or semi-public institutions operating at the Québec level offer some form of financial support, mostly in the form of investments, to technological initiatives in Gatineau. The *Caisse de dépôt et de placements du Québec* (Québec's public pension fund) and the *Fonds de solidarité du Québec* (a fund created by a Québec worker's union federation to invest in business) through its regional office have all provided significant venture capital for firms operating in

---

13 This is not to suggest that firms are not involved in the governance of the sector.

14 "The boom of the high tech sector is only taking place on the Ottawa side of the river, mainly in the Western part of the city. So far, Gatineau has been less successful in attracting these types of industries, as much of the expansion of technology-based clusters occurred through the 1990s on the Ontario side of this region. Such a dynamic provides testament to the importance of proximity in reinforcing the clustering of firms within specific territories (despite, in the case of Québec, more generous research and development incentives…)" (Paquet, Roy and Wilson, 2004, p. 230).

Gatineau. To these initiatives, one can add *Investissement Québec*'s important guarantees to back a firm's quest for private loans.

The federal government also proved to be an important player through its regional development policy for Québec. Canada Economic Development's (CED) Outaouais office has been very active in many aspects of HT development in Gatineau. It offers venture capital to Small and Medium Enterprises (SME) and has also invested in a certain number of projects which serve as incubators supporting the creation of new enterprises in Gatineau, notably the *Centre de développement des entreprises technologiques* (CDET) which are to be presented later on.

These provincial institutions as well as federal and regional support tend to reassert the relevance of the provincial border within the cluster. They offer resources (mainly financial) and support which are restricted to actors operating in the Outaouais region. As such, they can become alternative networks possibly supplying information and financial support that can counterbalance somewhat the pull from Ottawa.

Local Governance

The support offered by upper level governments and pan-Québec social actors is matched by the work of regional and local institutions and is very much intertwined in a complex local governance. Since at least the early 1990s, in the wake of a major rollback of federal government employment, many initiatives aimed at supporting local new economy growth have been initiated by these institutional actors (Doucet, Favreau and Robitaille, 2005). These initiatives often take the form of complex partnerships between public actors such as the City of Gatineau and its *Corporation de développement économique* or the rural *Municipalités régionales de comtés* surrounding Gatineau and semi-public institutions such as the *Conseil régional de développement de l'Outaouais*[15] and the *Centres locaux de développement* (local development corporations). These institutional partners organised at the local level rely quite heavily on funding coming from above (regional development or technological development funds).

Although they are never fully insulated from the dynamics of the Ottawa cluster (and often rely to a certain extent on the resources offered by it), these projects aim precisely at the promotion of the dynamism of

---

15 The *Conseil régional de développement de l'Outaouais* (CRDO) is a regional assembly which has the mandate to manage regional development funds coming from the Québec government for the region. The CRDO has identified technological development as one of its top priorities. Since 2003 and the change in government, the CRDO is now called the *Conférence régionale des élus de l'Outaouais*.

Gatineau's HT sector. As we will see, these initiatives contribute at two levels: 1. providing venture capital directly to existing and emerging firms and 2. promoting research and business development infrastructure. Two of these initiatives are briefly presented here in order to better gauge institutional involvement in local governance: the *Centre de développement d'entreprises technologiques* and the *Language Technologies Research Centre*.

Centre de Développement d'Entreprises Technologiques (*CDET*)

The CDET has served as a technological incubator for the region since 1989. Its principal objective is to offer professional services in order to facilitate the creation of new businesses. The services offered to entrepreneurs are manifold: entrepreneurial coaching (business plan, financial planning, project viability assessment, market strategies etc.), technical assistance (networking, technical validation) and marketing. They tend to concentrate on the initial phases of SME business creation (CDET, undated).

The services offered by the Centre seem to be in demand since more than 450 projects have been processed in the last seven years (1997-2004). Almost half of these projects were in the information technology subsector which is not surprising considering the profile of most current firms in Gatineau. However, it is interesting to note that last year the Centre received 22 demands concerning projects involving technological applications to forest product transformation (CDET, undated). These projects are particularly relevant considering the important role that the wood industry has traditionally played in the region and the urgent need for this industry to find more innovative products and new market niches in a technological sub-sector not invested and dominated by Ottawa firms.

The Centre is funded mainly by a federal partner (Canada Economic Development) as well as provincial (the Department of Economic Development, Innovation and Exports, *Ministère du développement économique, de l'innovation et de l'exportation*), regional (*CRDO*) and local (City of Gatineau) partners. Its board of directors is governed mainly by representatives from private firms (5), as well as from the City (2). It therefore embodies a form of associative governance based on cooperation between public and private actors from the Québec side (Paquet, Roy and Wilson, 2004). It has also developed an important link with the National Research Council (NRC) in Ottawa. The CDET has hosted – for more than ten years – the NRC's Industrial Research Assistance Programme.

Language Technologies Research Centre (LTRC)

The LTRC is a more recent initiative officially launched by the federal government's *Action Plan on Official Bilinguilism* in 2003 (Intergovernmental affairs, undated). In this Plan, the government expressed its desire to promote Canada as a world leader in language technologies (translation and terminology software, language training technologies and multilingual technologies) and announced a multimillion dollar investment in a Centre to be situated on the campus of the *Université du Québec en Outaouais*, in Gatineau. This new R&D lab is now under construction and allows collaboration between professors from the university specialising in linguistics and translation and researchers from the NRC who also to be on-site. It is expected that the lab is to accommodate up to 150 researchers (CRTL, undated). The Centre is also to house an incubator facilitating the creation of new businesses in the field of language technologies as well as promoting the use and marketisation of the technological products developed by internal researchers.

In order to convince the federal government to locate the language technology lab in Gatineau, much groundwork was laid by institutional actors operating in Gatineau. The University, partially through its outreach bureau, the *Conseil régional de développement de l'Outaouais* and the *Corporation de développement économique de Gatineau* all contributed significant resources, professional and otherwise, and energies in order to construct and promote the project. They secured the support and the collaboration of the Translation Bureau, a federal government branch located in Gatineau as well as the Language Industry Association. The project also capitalised on important financial support from regional development apparatuses from both levels of government: Canada Economic Development for the federal government and Québec's *Ministère du développement économique et régional* (CRTL, undated).

As we can see, the LTRC was constructed as a fairly complex partnerships between the university and actors from the public as well as the private sector. The project also has a definite cross-border dimension since the collaboration between institutions operating in Gatineau managed to secure the collaboration and some important research capacity from the federal government's NRC. This project is very significant for Gatineau, firstly because of the sheer size of the investment in R&D and the important number of researchers involved. Its significance also comes from the fact that it represents an R&D infrastructure in a technology subsector *not* dominated by Ottawa (Materazzi, 2004, p. 248). We can assume that in the mind of regional actors, language technologies is hoped to become a HT subfield were Gatineau has definite leadership.

Conclusions

This chapter's initial intent was to contribute to a better understanding of Gatineau's HT sectors and its integration/demarcation in the Ottawa cluster. Our analysis was concerned with two levels: the level of firms and the level of public (and semi-public) institutions. We have suggested that Gatineau firms are in many ways very integrated in the cluster. This integration can sometimes be very asymmetrical as is the case for most Gatineau businesses still predominantly relying on Ottawa R&D facilities. This integration appears much less clearly when one looks at the dynamics of institutions. As this chapter suggests, the Gatineau HT sector can rely on a variety of public as well as civil society institutions (unions and business associations) whose mandate is to specifically promote Gatineau's HT sector. These (regional and state) institutions cooperate in a complex web of partnerships which serve as alternative networks offering a variety of resources and expertise for local firms. The findings from the 2002 survey referred to earlier suggest that these resources (notably the financial resources but not exclusively) were important to ensure the survival of many firms:

> The start-up of these companies can be attributed to a number of factors. One is the growing presence of venture capital groups in the region since the early 1990s. In fact, among the 29 enterprises studied, half were financed by venture capital groups, primarily the Fonds régional de solidarité Outaouais (a labor-sponsored fund), the Société de diversification économique de l'Outaouais (SDEO, or Outaouais Economic Diversification Corporation), Cap-Gest-Dev, the Caisse de Dépôt et de placements du Québec (Québec's government run pension-fund manager), and Innovatech. The majority of these companies have received several rounds of financing and are looking for still more capital. Most have backing from at least 2 investment institutions, and a few have 3 or 4 backers. The government programmes such as Canada Economic Development for Québec Regions and Investissement-Québec, especially the loan guarantees to banks, are extremely useful….The Technology Business Development Centre has also incubated a good number of these start-up companies (Materazzi, 2004, p. 245).

The work of these HT-concerned institutions is interesting from the perspective of the border. Of course, these institutions have territories which are defined by the provincial border.

They identify with these territories as opposed to the Ottawa clustered region and their work and mandate is to ensure the development within the boundaries of these institutional spaces. This does not mean that they do not have links with firms and institutions across the river. In the case of the *Centre de développement des entreprises technologiques* as well as in the LTRC, important and probably very beneficial partnerships were established with research labs located in Ottawa. However, these partnerships were always realised with the specific aim to promote firms from the Québec side of the river. On a more theoretical note, in the confrontation between "spaces of flows" and "spaces of places," local institutions do not disappear or become irrelevant – far from it. However, they cannot function in a "command and control" fashion as they attempted in the past (Pal, 2002). As this chapter suggests, institutions have to find ways to adapt and channel the pressures coming from networks and functional space that operate on scales beyond and across their borders.

References

Andrew, C., Houle, F., Thériault, J. (1992). Le local dans les nouvelles stratégies de développement. *Canadian Journal of Regional Science, XV*, (3), 457-472.

Aydalot, P. (ed.). (1986). *Milieux innovateurs en Europe*. Paris: GREMI.

Badie, B. (1995). *La fin des territoires*. Paris: Fayard.

Beccatini, G. (1992). Le district marshallien: une notion socio-économique. In Benko, G. and Alain Lipietz. (eds.). *Les régions qui gagnent: Districts et réseaux. Les nouveaux paradigmes de la géographie économique*. Paris: Presses universitaires de France, 35-56.

Biarez, S. (1997). Métropolisation contre gouvernement métropolitain. In G. Saez, J.-P. Leresche and M. Bassand (eds.). *Gouvernance métropolitaine et transfrontalière: action publique territoriale*. Paris: L'Harmattan.

Blatter, J. (2004). 'From spaces of place' to 'spaces of flows'? Territorial and functional governance in cross-border regions in Europe and North America. *International Journal of Urban and Regional Research, 28* (3), 530-548.

Castells, M. (1996). *The rise of the network society*. London: Blackwell.

Centre de développement d'entreprises technologiques. (undated). *Historique du centre*. Gatineau.

CRTL (Centre de recherche en technologie langagière). *CRTL en bref.* URL address: http://www.crtl.ca/fr/bref.htm.

Chaiton, A., Dibbits, T., Paquet, G., Roy, J., Wilson, C. (2000). *The collaborative challenges of regional governance. Innovation, new social technologies and civic entrepreneurs*. Paper presented at the Knowledge Based Economy Workshop. Vancouver: December.

Conseil de la science et de la technologie (2001). *Rapport de conjoncture 2001. Pour des régions innovantes*. Québec: Government of Québec.

Doucet, C., Favreau, L., Robitaille, M. (2005). *Portrait économique de l'Outaouais et de ses régions*. Gatineau: Chaire de recherche sur le développement des collectivités.

FM Consult. (2002). *Portrait technologique de l'Outaouais*. Gatineau.

Fontan, J-M., Klein, J.-L., Tremblay, D.-G. (eds.). (1999). *Entre la métropolisation et le village global*. Québec: Presses de l'Université du Québec.

Gaffield, C. (ed.). (1996). *Histoire de l'Outaouais*. Québec: Institut québécois de recherche sur la culture.

Godin, B. (2000). La mesure de la science et la construction statistique d'un territoire: la région de la capitale nationale du Canada. *Canadian Journal of Political Science, XXXIII* (2), 333-359.

Harrison, R.T., Cooper, S.Y., Mason, C.M. (2004). Entrepreneurial activity and the dynamics of technology-based cluster development: the case of Ottawa. *Urban Studies, 41* (5/6), 1045-1070.

ICF Consulting. (2000). *Choosing a future. A new economic vision for Ottawa*. Ottawa: ICF.

Intergovernmental Affairs. (undated). URL address: http://www.pco-bcp.gc.ca/aia/ default.asp?Language=F&Page=ActionPlan.

Lipietz, A., Leborgne, D. (1992). Flexibilité offensive, flexibilité défensive. Deux stratégies sociales dans la production des nouveaux espaces économiques. In Benko, G. and A. Lipietz (eds.). *Les régions qui gagnent: Districts et réseaux. Les nouveaux paradigmes de la géographie économique*. Paris: Presses universitaires de France. 347-377.

May, N. (1986). Constitution d'un regard. Fordisme et localisme. *Les annales de la recherche urbaine, 29*, 5-12.

Madill, J., Haines, G., Riding, A. (2004). A tale of one city: the Ottawa technology cluster. In L. Shavinina (ed.). *Silicon Valley North. A High-tech Cluster of Innovation and Entrepreneurship*. Amsterdam, Elsevier. 85-117.

Maillat, D. (1992). Milieux et dynamiques territoriales de l'innovation. *Canadian Journal of Regional Science, XV* (2), 199-218.

Pal, L. (2002). *Beyond policy analysis: public issue management in turbulent times*. Scarborough: Nelson.

Paquet, G., Roy, J., Wilson, C. (2004). Ottawa's TalentWorks-regional learning and collaborative governance for a knowledge age. In L. Shavinina (ed.) *Silicon valley north. A high-tech cluster of innovation and entrepreneurship*. Amsterdam: Elsevier. 311-329.

Pecqueur, B., Rousier, N. (1992). Les districts technologiques, un nouveau concept pour l'étude des relations technologies-territoires. *Canadian Journal of Regional Science, XV* (3), 437-456.

Piore, M., Sabel, C. (1984). *The second industrial divide. Possibilities for prosperity*. New York: Basic Books.

Proulx, M-U. (1992). Innovative milieus and regional development. *Canadian Journal of Regional Science, XV* (2), 149-154.

Porter, M. (1990). *The competitive advantages of nations*. New York: Free Press.

Saxenian, A. (1994). *Regional advantage. Culture and competition in Silicon Valley and Route 128*. Cambridge: Harvard University Press.

Scott, A., Storper, M. (1992). *Production, work, territory: the geographical anatomy of industrial capitalism*. Boston: Allen and Unwin.

SDEO. (undated). *La société de diversification économique de l'Outaouais*. Gatineau.

Shavinina, L. (2004). Silicon phenomenon: introduction to some important issues. In L. Shavinina (ed.) *Silicon Valley North. A high-tech cluster of innovation and entrepreneurship*. Amsterdam: Elsevier. 3-19.

Storper, M. (1997). *The regional world: territorial development in a global economy*. New York: Guilford Press.

Tremblay, D-G. (1998). Partenariat, coopération et imbrications locales: une analyse des limites et des perspectives fondées sur les résultats d'une enquête menée au Québec. In Marc-Urbain Proulx (ed.). *Territoires et développement économique*. Montréal: L'Harmattan.

Tremblay, R. (2004). Ottawa-Gatineau et le savoir. *Organisations & Territoires*, *13* (1), 65-70.

Wolfe, D., Holbrook, A. (ed.). (2000). *Innovation, institutions and territory. Regional innovation systems in Canada*. Montréal: McGill-Queen's University.

Wolfe, D. (2003). From the national to the local. Recent lessons from economic development policy. In Caroline Andrew, Katherine Graham and Susan Philips (eds.). *Urban Affairs. Back on the Policy Agenda*. Montréal: McGill-Queen's University Press, 239-264.

Annex

Institutions Included in the Study and Their Websites

Développement économique Canada

- Agence de développement économique du Canada pour les régions du Québec. (Consulted in May 2005). Address: http://www.decced.gc.ca/asp/SalleNouvelles/communiques.asp?LANG=FR&TYPE_COMM=COMM&PRESS_RELEASE_ID=3976.
- Agence de développement économique du Canada pour les régions du Québec. (Page consultée en mai 2005). Initiative régionale stratégique (IRS) de l'Outaouais, [En ligne]. URL address: http://www.decced.gc.ca/asp/ProgrammesServices/irs_outaouais.asp?LANG=FR.

Association de la recherche industrielle du Québec

- Association de la recherche industrielle du Québec. (Consulted in May 2005). PME Innovation. Address: http://www.adriq.com/PME_Innovation/index.html.

Caisse de dépôt et de placement

- Caisse de dépôt et de placement. (Consulted in May 2005). CDP Capital-Technologies. Address: http://www.cdpcapital.com/ La-Caisse/Filiale/fr/ HTML/1945_fr.aspx.

Centre de développement d'entreprises technologiques

- Centre de développement d'entreprises technologiques. (Consulted in May 2005). Services intégrés au centre. Address: http://www.cdet.ca/cdetrev02_03/cdetf.htm.

Centre local de développement de Gatineau

- Centre local de développement Gatineau. (Consulted in May 2005). Politique d'investissement du « Fonds local d'investissement » « FLI » du CLD Gatineau. Address: http://www.cldgatineau.ca/documents/politique-fli.pdf.

Corporation de développement économique de Gatineau

- Corporation de développement économique de la Ville de Gatineau. (Consulted in May, 2005). Investir à Gatineau, [Online]. URL address: http://www.directiongatineau.ca/investir.htm.
- Corporation de développement économique de la Ville de Gatineau. (Consulted in May, 2005). Secteurs d'intervention, [Online]. URL address: http://www.directiongatineau.ca/secteurs.htm.

Fonds régional de solidarité Outaouais

- Fonds régional de solidarité Outaouais. (Consulted in May, 2005). Modalités d'intervention. URL address: http://www.out.fondsreg.com/frsout.nsf.

CHAPTER 11

Immigrants in Ottawa's Information Technology Sector

A Profile

Amelia BROWN

Fulbright Foundation

Introduction

A largely untold, but important, part of the story of the dramatic rise and subsequent downturn of the information technology sector in Ottawa is immigration. Although foreign-born workers accounted for a large proportion of the technology workforce during the boom and an even larger proportion of those unemployed in the bust, very little is known about their role in the technology sector in Ottawa, or, for that matter, in Canada as a whole. Even basic data on how many immigrants work in the technology sector are difficult to come by. Beyond the few statistics available at the national level based on the census (see Habtu, 2003; Gagnon *et al.*, 2003), a handful of studies (Rao, 2001; Malkam Cross-Cultural Training, 2000), and occasional mention in local newspaper stories or presentations by business leaders, information on foreign-born technology workers in Canada is extremely limited. Information on immigrants in Ottawa's technology sector is even less available.

This chapter is meant to be a small step toward addressing this knowledge gap. Understanding more about the role of immigrants in the technology sector in Canada is important not only for the sake of immigrants but also for society as a whole, enabling decision makers and corporate leaders to design policies and practices that ensure that these newcomers' skills do not go to waste. Information on immigrants in Ottawa's technology sector is useful for local residents, companies, and officials, but it also adds to the general knowledge base on technology workers in Canada.

This chapter is divided into five broad sections. First, to provide a larger context for the Ottawa discussion, the chapter provides an overview of the technology sector and technology workforce in Canada. Second, the chapter looks at immigrants in Canada's technology sector. Third, the chapter presents background information on the technology sector in Ontario and Ottawa and examines immigration to Ottawa in technology occupations. Fourth, the chapter looks at the prominent role of Chinese immigration in Ottawa's technology sector and presents a small case study of skilled Chinese immigrants in technology occupations in Ottawa. Finally, the chapter concludes by offering some observations and recommendations based on the data presented and the findings of the case study.

The statistics presented in this chapter are drawn from various sources. All of the information on immigration flows to Canada and Ottawa in technology occupations, in the text and in the tables and charts, was specially generated by Citizenship and Immigration Canada for this study. Other data are drawn from secondary sources. The types of measurements and the definitions of the technology sector and the technology workforce vary from source to source, making comparisons across studies difficult. Where applicable, I have noted in the endnotes the definition or measurement used. While the diverse sources presented are not uniform in their approach, they offer, when viewed as a whole, a good general picture of the technology sector and workforce. Throughout the chapter, 'Ottawa' refers to the Ottawa-Gatineau census metropolitan area.

Canada's Information Technology Sector and Workforce

The information and communications technology (ICT) sector[1] in Canada has experienced phenomenal growth over the past decade, even in light of the economic downturn in the sector in recent years. During the 1990s, employment in the sector increased by over 70%, a much higher percentage than in comparable skill-intensive sectors (Beckstead *et al.*, 2003, p. 6). Between 1997 and 2000, ICT sector growth was five times the rate of growth of the total economy (Science, Innovation and

---

1 Throughout the chapter, "ICT sector" refers specifically to the information and communications technology sector as defined by the Organisation for Economic Co-operation and Development (OECD). The terms "technology sector," "high-tech sector," and "information technology sector" are used to refer to the sector in a more general sense.

Electronic Information Division, 2001, p. 10). The economic downturn in the ICT sector, which began in 2000, hampered this growth and led to many job losses. Revenues from the sector dropped by 10% in two years due primarily to a dramatic decrease in revenues in ICT manufacturing industries (ICT Branch, 2004). Despite the downturn, the ICT sector has remained an important part of the Canadian economy. The sector accounted for 57 billion Cdn$[2] of Canada's gross domestic product (GDP) in 2004 (5.4% of total GDP), representing an increase of 4.1% over the year. More than 10% of national growth since 1997 stems from the ICT sector. The ICT sector is also a major contributor to research and development (R&D), with estimated R&D expenditures in the ICT sector representing 40% of total Canadian private sector R&D spending in 2004 (ICT Branch, 2005).

ICT industries have rebounded to some degree since the worst of the downturn in 2001, and observers have been forecasting a continued recovery. In a recent report, the Conference Board of Canada predicted that profits in the sector would reach a five-year high in 2005 and continue to rise in 2006 (Burt, 2005). Almost three-quarters of CEOs surveyed by consulting firm Deloitte from among North America's fastest growing technology companies are very or extremely confident that their high growth rates are to continue in the next year, and the vast majority plan to increase staff in the coming year (Deloitte, 2005). Furthermore, a recent Statistics Canada study of the high-tech boom and bust concluded that the downturn was a temporary phenomenon, rather than part of a structural shift leading to long-term slow growth (Beckstead and Brown, 2005).

How have information and technology workers fared over the past few years? Overall, from 1997 to 2003, the ICT sector workforce increased by 27.8%, with an annual growth rate almost double that for the economy as a whole. After stagnating in 2001 and falling in 2002, the number of persons employed in the ICT sector grew by over 5% in 2003, although still not reaching the level of employment at the height of the boom. Workers in the ICT sector represented 3.6% of all Canadian workers in 2003 (ICT Branch, 2005). Not all technology workers, however, are employed in the ICT sector, and not all workers in the ICT sector are in high-technology occupations. The majority of workers in high-technology occupations are in fact employed in sectors other than the ICT sector (O'Grady, 2004).

For a complete picture of the information technology labour force, one must look at workers by occupation category, rather than by sector.

2 In 1997 dollars.

Employment in information technology occupations nearly doubled from the time of the 1991 Census to the 2001 Census, both in terms of raw numbers and proportionately, increasing from 1.8% of the overall Canadian labour force in 1991 to 3.3% in 2001 (Gagnon *et al.*, 2003, p. 16).[3] According to data from Statistics Canada's monthly Labour Force Survey, the information technology labour force has remained close to half a million from the end of 2000 to the end of 2004, with a low of 485,000 in late 2001 and a high of 516,000 in August 2002 and January 2003.

The downturn did not lead to dramatic reductions in the technology workforce, but it did halt what had been a pattern of rapid growth in the workforce. The downturn also led to higher unemployment rates for technology workers. At the end of 2000, the unemployment rate among technology workers was 2%. The rate increased steadily to reach 5.5% in the fall of 2001. From late 2001 to mid-2004, the unemployment rate fluctuated between just below 4% and just above 5.5%. From mid-2003 to late 2004, the general trend was downward, with unemployment at 2.8% in December 2004 (Wolfson, 2005, p. 4-6).[4]

Who makes up the information technology workforce? Based on the Labour Force Survey data from 2000 to 2004, the large majority of workers, 78%, are male. Most are young, with 45% of workers under age thirty-five and 34% of workers between ages thirty-five and forty-four. Three-quarters of the labour force have a post-secondary degree. Almost all workers hold full-time, permanent jobs (Wolfson, 2005, p. 9-13). Technology workers also have relatively high incomes. At 47,709 Cdn$, average earnings for technology workers in 2000 were 1.5 times higher than for employed Canadians as a whole (Gagnon *et al.*, 2003).

3 1991 and 2001 Census data for information technology workers include twelve occupations under the 2001 National Occupational Classifications System (NOC). 1991 occupation codes have been matched with equivalent 2001 codes. See Gagnon *et al.* (2003) for an exact listing of occupation codes.

4 The data on the information technology labour force from the Wolfson report refer to twenty-one occupations within the National Occupational Classifications System (NOC) identified by Human Resources and Skills Development Canada and the Software Human Resource Council. See Wolfson 2005 for a complete listing of occupation codes.

Foreign-born Technology Workers in Canada

According to census data, 32% of technology workers[5] in 2001 were foreign-born. The proportion of immigrants was even higher in certain technology occupations: 47% of software engineers, 40% of computer engineers, and 37% of computer programmers were foreign-born in 2001. Immigrants are much more prevalent in technology occupations than in the economy more generally. In 2001, immigrants accounted for only 20% of workers in all occupations (Habtu, 2003).

A large proportion of foreign-born technology workers are recent immigrants. Nearly half of technology workers who were born outside of Canada immigrated in the 1990s, and one-third arrived after 1996. Immigrants working in technology occupations are more likely to have arrived recently than immigrants working in occupations across all categories. Less than one-third of foreign-born workers in Canada as a whole arrived in the 1990s (Habtu, 2003).

The share of technology workers born abroad has increased over time. In 1961, 24.4% of the information technology labour force was foreign-born, a proportion that increased only slightly in the 1970s and 1980s. The proportion of immigrants in the information technology sector rose substantially in the 1990s, however, coinciding with the growth of the technology sector. The proportion of immigrants in technology occupations has increased steadily since the 1960s relative to the proportion of immigrants in the labour force as a whole. In 1961, the difference between the proportion of immigrants in the information technology workforce and the general workforce was 2.7% (with a higher proportion in technology occupations than in overall occupations); in 1981, the difference was 7.0%; and in 2001, the difference was 11.1% (Gagnon *et al.*, 2003, p. 25).

With immigrants accounting for a substantial portion of the technology workforce, particularly during the rapid growth of the technology sector in the 1990s, immigration has clearly played a significant role in the development of technology industries in Canada. This is underscored by the fact that immigrants make up an especially large proportion of workers in some of the highest-skilled technology occupations, such as computer engineer and computer programmer – occupations that drive the innovation that has been a prime source in the dramatic growth of the technology sector.

---

5 Habtu (2003) includes nine occupation categories under the natural and applied sciences and related occupations group in her definition of the technology workforce. See Habtu 2003 for a complete list of NOC codes used.

The increasing role of immigrants in technology occupations is also apparent in immigration statistics. The number of skilled immigrants coming to Canada to work in technology occupations[6] has risen dramatically with the growth of the high-tech sector in Canada (Table 1). A total of 876 individuals in technology occupations landed in Canada in 1990. In 2002, over 14,000 technology workers immigrated to Canada (17,293 in 2001), representing 27% of the total number of skilled workers to Canada. China has dominated the growth in technology workers, accounting for 28% of the total tech immigration from 1990 to 2002. India is in second place, accounting for 9% of total tech immigration from 1990 to 2002.

These large numbers, in part, reflect Canada's immigration policy. Attracting skilled immigrants has been a primary goal of Canada's immigration policies in recent years, and Canada has become a world leader in the development of policies to encourage the immigration of highly skilled workers. In particular, the Canadian points system, which admits permanent immigrants by awarding numerical scores based on their skills and background, has served as a model for many other countries. In addition to its permanent immigration system, Canada also admits foreign skilled workers through its temporary immigration system.[7] This chapter focuses primarily on permanent immigrants, and the immigration data presented include only permanent entries. However, temporary workers who were residing in Canada at the time of the 2001 Census are included in the census data on foreign-born technology workers.

Canada admits skilled workers through its permanent immigration system mainly through the skilled worker category.[8] Since the late

---

6 The numbers in this paper on the immigration of technology workers to Canada and Ottawa are from CIC's Landed Immigrant Data System. They include only principal applicants under the skilled worker category who listed one of the following occupations as their intended job: computer programmer (NOC 2163), computer systems analyst (NOC 2162), computer engineer (including hardware and software engineers) (NOC 2147), and electrical and electronics engineer (NOC 2133). Participants in the Software Pilot Project who applied for landed status were also included. All occupations were in skill level A (occupations usually requiring a university education) or 0 (management occupations).

7 In 2003, Canada admitted 82,151 foreign workers through its temporary immigration system, 1,867 of whom were destined to Ottawa. Of these, 17,142 were in managerial occupations or those generally requiring a university degree (skill level 0 and A). As of 1 December 2003, a total of 111,195 temporary foreign workers were residing in Canada, 2,849 in Ottawa. 37,146 of these residents were in jobs rated as skill level 0 or A (CIC, 2004).

8 Immigrants who are highly skilled also enter through other streams that are not skills based such as the family class (spouses or close relatives of Canadian citizens or

1960s, with the introduction of the points system, Canada's permanent immigration programme has focused heavily on the skills of incoming immigrants (Green and Green, 1996, p. 18-22). Recent years have seen a redesign of the points system to focus less on short-term economic needs in specific occupations and more on the need for general skills relevant to long-term economic growth.

Under the current points system, individuals who apply for permanent residence through the skilled worker category receive points according to various selection criteria: education, official languages, employment experience, age, arranged employment in Canada, and adaptability. The maximum number of points is 100, and 67 points are currently needed to pass the skills test. If an individual passes the skills test, can demonstrate that they have the financial means to support themselves in Canada, and meets other general migration requirements (such as good health, no criminal record), then they are admitted to Canada as a skilled worker. The entire process can be as short as one year, but more often takes longer.

In 2003, 105,240 immigrants entered Canada under the skilled worker category, 45,378 of who were principal applicants.[9] Skilled workers and their dependents accounted for 48% of total immigration to Canada in 2003, up from 31% in 1994 (CIC, 2004). The recruitment of workers in very high-skill occupations, such as information technology occupations, has been an important part of Canada's strategy to attract skilled immigrants.

Along with the rise in skilled immigration overall throughout the 1990s, the occupational skill level of immigrants has increased. In 1990, for those immigrants for whom occupation skill levels were listed, 17% were in skill level A, professional jobs generally requiring a university education; in 2000, 60% of the occupations were in skill level A. Among principal applicants in the skilled worker category, computer programmer and computer systems analyst were the two most common occupations over the 1996 to 2000 period, and 20% of skilled worker principal applicants were in one of these two occupations in the year 2000 (CIC, 2003a).

---

permanent residents), as refugees or asylees, or through special programs such as the Provincial Nominee Programme. Canada also has a business immigrant class, through which persons willing to invest substantially in Canada are able to obtain permanent residence.

9 Principal applicants are those who completed the application to come to Canada as skilled workers and will enter the labour force as employees, as opposed to dependents of principal applicants who are also included in the skilled worker category.

Skilled Immigrants and High Technology in Ottawa, Ontario

Ontario is without question the centre of the technology sector in Canada. In 2002, Ontario accounted for 49% of the ICT sector's GDP and 55% of its exports. Over two-thirds of the sector's research and development spending (4.6 billion Cdn$ in total) was concentrated in Ontario in 2001 (ICT Branch, 2004). During the 2000-2004 period, one-half of all workers in information technology occupations were in Ontario (Wolfson, 2004, p. 12). Ontario also has by far the largest proportion of workers in technology occupations, with 4.4% of all workers in the province employed in technology occupations in 2001 (the next highest proportion, 3.2%, was in Québec). Technology workers in Ontario have higher employment earnings than their counterparts in other provinces (Jacobs, Vaillancourt, and Vaillancourt, 2004, p. 6-8).[10]

Within Ontario and the country as a whole, Ottawa stands out as a leader in technology. While Toronto has the largest number of workers in the information technology sector, the Ottawa metropolitan area has the highest proportion of workers in information technology out of any city in Canada. In 2000, 9% of all workers in Ottawa were employed in the ICT sector,[11] an increase of 61% from 1990. In absolute numbers, Ottawa's technology workforce ranks fourth in the country (Beckstead *et al.*, 2003, p. 24-27).[12] Due to its strong technology sector, Ottawa is also a national leader in research and development. In 2001, Ottawa had the most patents granted per capita of any major Canadian city (Centre on Governance, 2002b, p. 25).

Estimates of the exact number of technology workers in Ottawa vary, depending on the measurement used. According to the 2001 Census, the Ottawa metropolitan area was home to 50,880 technology workers, as compared to a technology workforce of 147,615 in Toronto (Jacobs, Vaillancourt, and Vaillancourt, 2004, p. 2). The Centre on

---

10 Data in Jacobs, Vaillancourt, and Vaillancourt 2004 are from the 2001 Census and include twelve information technology occupations under the 2001 National Occupational Classifications System (NOC).

11 Other estimates, using slightly different definitions of the information technology sector, place the percentage even higher. The Centre on Governance estimated that 13% of Ottawa's workforce was in the technology sector in 2001 or 73,900 people in total (Centre on Governance, Report I, 2002, p. 39).

12 Employment data in Beckstead *et al.* (2003) are for the ICT sector, as defined by the OECD, and are drawn from Statistics Canada's Business Register and Labour Force Survey.

Governance, using a wider definition of technology workers, estimated one hundred thousand technology workers in the national capital region for the same year (Centre on Governance, 2002b, p. 26). A more recent tabulation by the Ottawa Centre for Research and Innovation (OCRI) places the size of the workforce at 67,800 as of January 2005 (OCRI, 2005). Within Ottawa's economy, the technology sector represents the second largest industry sector in terms of employment after the public sector (Centre on Governance, 2002b, 26).

The technology sector in Ottawa has mirrored the economic pattern of the technology sector at the national level, experiencing rapid growth in the 1990s, followed by an economic downturn. According to Statistics Canada figures, technology employment in Ottawa peaked in May 2000 at 72,400 and then generally declined (with ups and downs along the way), reaching an eight-year low in December 2004 of 41,800. The first few months of 2005 have seen strong job growth in the sector, with gains of almost 7,000 jobs in January and February (Hill, 2005). Many are predicting continued growth in Ottawa's technology sector, but forecasts vary. OCRI president and CEO Jeffrey Dale has been most optimistic, foreseeing an additional 30,000 net jobs added in the next three years (Sutcliffe, 2005). Most observers, however, are predicting much more modest growth.

Exact figures on the proportion of immigrants among Ottawa's technology workforce are not readily available, but statistics would likely be similar to those at the national level. Data on the entry of immigrants in technology occupations who intend to settle in Ottawa suggest that immigrants are a significant part of Ottawa's technology sector. Just over 5,700 skilled immigrants in technology occupations (computer programmer, computer systems analyst, computer engineer, or electrical/ electronic engineer) made Ottawa[13] their destination from 1990 to 2002. These persons account for a significant proportion of total skilled immigration to Ottawa.

Of the 18,605 principal applicants in the skilled worker category who listed Ottawa as their destination from 1990 to 2002, nearly one-third intended to work in information technology occupations. The

[13] In this chapter, the numbers on the immigration of technology workers to Ottawa are from CIC's Landed Immigrant Data System. Ottawa is defined as the Ottawa-Gatineau metropolitan area: Blackburn Hamlet, Carp, Cumberland, Gloucester, Greely, Kanata, Kinburn, Manotick, Navan, Nepean, North Gower, Orleans, Osgoode, Ottawa, Richmond, Rockliffe Park, Rockland, Stittsville, and Vanier in Ontario and Angers, Aylmer, Buckingham, Cantley, Chelsea, Deschenes, Gatineau, Hull, Luskville, Masson, Perkins, Pointe Gatineau, Quyon, Ste-Cecile-de-Masham, Templeton, and Wakefield in Québec.

proportion is even higher if one looks at recent years alone: 44% of incoming skilled immigrants to Ottawa from 2000 to 2002 intended to work in high tech. Furthermore, among the top twenty occupations for skilled workers coming to Ottawa from 1990 to 2002, computer programmer is at the top of the list. Electrical and electronics engineer, computer systems analyst, and software engineer rank third, fourth, and eighth, respectively (Table 2).

As the information technology sector has grown in Ottawa, so too have the number of immigrants to Ottawa intending to work in the tech sector. In 1990, only 54 immigrants in highly skilled technology occupations listed Ottawa as their destination. By 2002, that number had risen to 736. At a peak in 2001, 1,037 immigrants in tech occupations landed in Ottawa (see Table 3 and Chart 1). A comparison of Ottawa's immigration numbers to total immigration to Canada illustrates Ottawa's status as a high-tech magnet. From 1990 to 2002, immigration of highly skilled technology workers to Ottawa represented 5.7% of all immigration of highly skilled technology workers to Canada. This is significantly higher than Ottawa's overall share of immigrants, which was 3.4% in the same period.

Statistics on the effects of the economic downturn on immigrants working in the technology sector are not available. Anecdotal evidence, however, suggests that immigrants were disproportionately affected. LASI (Local Agencies Serving Immigrants) World Skills, an Ottawa outreach organisation that assists immigrants with labour market issues, estimates that 80% of the technology workers who lost their jobs in Ottawa in recent years are immigrants (Centre on Governance, 2002b, p. 47). Furthermore, a large proportion of the technology workers in the 2002 case study that is the subject of this chapter were unemployed or engaged in activities outside of the technology sector.

That immigrants in technology occupations have been particularly impacted by the economic downturn is not surprising. In general in Ottawa, according to the 2001 Census, recent immigrants of working age with a university degree are more than four times as likely as Canadian-born persons with the same level of education to be unemployed (CIC, 2005a, p. 42). This is higher than the rate in Canada as a whole, where immigrants in the same category are slightly more than three times as likely as the Canadian-born to be unemployed (CIC, 2005b, p. 47). Furthermore, of university-educated recent immigrants who are employed in Ottawa, proportionately more work in jobs well below their skill level than do the Canadian-born. 17% of employed university graduates who immigrated to Canada between 1996 and 2001 worked in jobs that required less than a college degree or trade apprenticeship, as compared to only 9% of Canadian-born persons with university degrees (CIC, 2005a, p. 55).

In the wake of the tech downturn and the resulting tighter employment market, companies have been looking for more than solid technical skills. Hiring has focused much more on interpersonal skills, networks, and previous employment experience (Centre on Governance, 2002, p. 28-29). Given this trend, recent immigrants are at a particular disadvantage. Having newly arrived to Canada, they do not often have the Canadian job experience and personal connections that their Canadian-born counterparts possess. Limited language skills and unfamiliarity with Canadian culture can also limit the competitiveness of recent immigrants in the job market. Certain aspects of the Ottawa job market make finding employment especially difficult for immigrants. The public sector's dominance in Ottawa leads to additional requirements for many area jobs, such as security clearances (which can be difficult to obtain for persons who have resided outside of Canada for substantial periods of time), Canadian citizenship, and the ability to work in both French and English. Even jobs outside of the federal government may have these requirements if the company works closely with the federal government.

Skilled Chinese Immigrants in the High-tech Sector A Case Study

Recent Trends in Mainland Chinese Migration to Canada and to Ottawa

An estimated 30 million Chinese live outside of China. Three hundred thousand to four hundred thousand Chinese persons emigrate each year, with over one hundred thousand legally immigrating to the United States, Canada, or Australia annually. In Canada, the Chinese are the top source of skilled immigration – as well as the top source of all immigration. In the year 2002, 17.5% of principal applicants in the skilled worker category were Chinese, much higher than the second-ranked country, India, which accounted for 11% of skilled workers (CIC, 2003b, p. 97). In Ottawa, the Chinese are also the top source of both total immigration (see Table 4) and skilled immigration, representing 17% of all immigration and 18% of skilled workers in the year 2002.

The Chinese also dominate immigration in information technology occupations (see Table 3). Among highly skilled information technology workers who came under the skilled worker category in 2002 and listed Ottawa as their destination, over one-third were Chinese (down from a high of 48% in 2000). Since 1993, China has consistently been the largest source of skilled immigration to Ottawa in technology occupations. The growth in Chinese immigration to Ottawa in technology occupations has been dramatic. Although the numbers declined in 2002, in the five-year period preceding 2002, Chinese immigration to Ottawa in high-tech occupations increased by 255%. This is compared to a 93% increase in the same period for non-Chinese immigration of technology workers to Ottawa.

The magnitude of the growth in Chinese immigration in the area of information technology becomes even more evident when compared with immigration patterns for Indians, another group that has experienced significant growth in Ottawa and has consistently been ranked second since 1993 as a source of technology immigration to the city (see Charts 2, 3, 4). In the early 1990s, India and China sent similar numbers of immigrants in technology occupations to Ottawa. Since 1995, however, Chinese-born immigrants have far exceeded the numbers of Indian-born immigrants coming to work in Ottawa's high-tech sector.[14]

Skilled immigrants and those in technology occupations account for a large proportion of Chinese immigration to Ottawa. Among Chinese immigrants who landed in the past ten years and listed Ottawa as their destination, many more came under the skilled worker category (including dependents) than under the family category (see Chart 5). Nearly one-third of skilled immigrants (principal applicants) to Ottawa from China during the 1990s were in high-tech occupations. If one looks at just the recent period of 1998 to 2000, the annual average is 60%.

Case Study Methodology and Findings

In the spring of 2002, interviews were conducted with twenty-two recent Chinese immigrants to Ottawa in an effort to better understand the factors that influence the migration decisions of high-tech workers. The subject group was skilled immigrants from mainland China currently living in the Ottawa metropolitan area who landed in Canada as high-tech workers after 1994. Participants were identified using a snowball methodology. The interviews were structured, with the exception of a few open-ended questions, lasted approximately one hour, and were mostly conducted in Mandarin. Questions focused on the subject's

[14] It is interesting to note that India is the top sending country among temporary workers in Ottawa in the natural and applied sciences. On June 1, 2001, 300 Indians in Ottawa held work permits in the natural and applied sciences, as opposed to only 30 Chinese (Citizenship and Immigration Canada, Client-Based Data System).

life before moving to Canada and his or her reasons for leaving China; his or her reasons for choosing to come to Canada and to Ottawa in particular; life and job experiences in Ottawa to date and how these compare to previous expectations; and future plans.

Of the group interviewed, all had a bachelor's degree as a minimum before coming to Canada, and all but one were working in high-tech occupations in China. More than half were married, and about a third had children, before they moved. Economic motivations were the main drivers behind leaving China. Most of those interviewed listed better educational opportunities and job opportunities in Canada as the primary reasons for emigrating. Many also mentioned moving to improve the future for their families. Several indicated clean air and environment as important factors behind their decision.

When asked why they chose Canada in particular, in addition to the reasons just given, some individuals mentioned Canada's immigration system: that Canada was easier to immigrate to than many other countries and also that citizenship was easier to obtain in Canada. Several persons mentioned coming to Canada to obtain Canadian citizenship or job experience in order to return to China or move to another country. Only three of the respondents came to Canada to accompany or join family or friends. This is not entirely surprising, since most Chinese immigration to Ottawa is fairly recent, so extensive family networks are only beginning to form. Previous experience in Canada played no role in respondents' decisions. None had ever visited Canada before moving. Most had only limited knowledge of Canada before coming. What respondents did know about Canada they most commonly learned through the Internet, in keeping with their technology backgrounds. Others learned about Canada from friends or family already in Canada, from television, or from immigration consulting agencies. Close to half of those interviewed considered applying to other countries, primarily Australia and the United States, but none did.

None of the respondents had jobs arranged in Canada before immigrating. With most arriving during the height of the technology boom, all but three of the respondents were able to find work in Canada, and most were able to find full-time jobs in the high-tech sector. However, most of these jobs were short-lived, due largely to layoffs or companies going out of business. Almost half were able to find subsequent jobs, although not necessarily in the technology sector or in skilled occupations. Many were still unemployed or had left the labour force to pursue further schooling. One third of respondents said their overall job experience in Canada had not met their expectations or had fallen well below their expectations. Half felt that the experience had met their expectations, and two said the experience had somewhat exceeded their expectations.

Respondents mentioned several difficulties in looking for work. By far the most prevalent problem was language. Next most common was not knowing how to find a job and foreign job experience not being accepted in Canada. Several persons also mentioned a shortage in available jobs as being a problem, most likely an indication of the downturn in the high-tech sector.

In terms of respondents' future plans, none indicated staying in Canada longer than five years, although almost half were unsure of their future plans. Several mentioned that how long they stayed depended on their job prospects in Canada. Almost all respondents planned to become Canadian citizens or were already Canadian citizens. By far the most prevalent reason mentioned for becoming a Canadian citizen was to be able to travel more freely or to move to other countries. After China, the United States was the most common intended destination. Only one person mentioned becoming a Canadian citizen because they planned to settle permanently in Canada.

The small size of the case study limits the ability to draw firm conclusions about technology workers in general based on the findings. However, some overall observations from the case study, while not definitive, are still useful. The main conclusion is that economics was what was most important both in terms of attracting the workers to come to Canada and in terms of influencing their decisions to stay in Canada. The technology workers who were interviewed came for primarily economic reasons and were planning to stay in Canada only if doing so met their economic goals.

The relative ease with which citizenship can be obtained in Canada was also a major motivator for coming and a short-term motivator for staying (with most wanting to remain in the country long enough to obtain their Canadian citizenship). However, this too was part of an economic strategy, rather than representing a long-term commitment to Canada. The immigrants interviewed wanted to obtain Canadian citizenship because it would enhance their economic prospects in Canada, in China, and in other countries. Whether immigrants planned to stay in Canada for the long term depended on their job and earning opportunities in Canada.

Conclusions

This chapter has attempted to add to the general understanding of the role of immigrants in the technology sector in Ottawa and in Canada as a whole by bringing together existing information from secondary sources and providing new data. A few conclusions can be drawn from the information presented.

First and foremost, the chapter highlights the need for further information on foreign-born technology workers and their role in the technology sector. As previously mentioned, existing research in this area is extremely scarce. Some data is available through government sources, such as the census and Citizenship and Immigration Canada databases, but much of these are not publicly available. Groups like the Software Human Resource Council (SHRC) and the Ottawa Centre for Research and Innovation (OCRI) have been working with government departments to produce studies focusing on the technology sector, but immigration is often not part of these efforts. Having a complete picture of the relationship between immigration and the technology sector is the first step in being able to design effective policies in this area. The benefits of better understanding this relationship also extend to other, more general policy areas, such as skilled immigration policy, immigrant economic integration, and economic development and growth.

Furthermore, a discussion of the role of immigrants in the technology sector needs to be part of the dialogue on Ottawa's economic development. Over the past few years, the role of technology workers and the role of immigrants in the economy have become part of the community development discourse in Ottawa. OCRI and the Ottawa Talent Initiative, an organisation formed in 2004 to assist unemployed and underemployed technology workers in Ottawa, have been working to put the needs of technology workers on Ottawa's policy agenda.[15]

Several organisations and projects have also been attempting to improve the integration of immigrants into Ottawa's labour market.[16] However, the two discussions, one related to technology workers and the other related to skilled immigrants, remain largely distinct. The challenge is to integrate these two discussions by putting the role of immigrants in the technology sector on the agenda. As the data in this chapter show, technology workers make up a substantial proportion of Ottawa's immigrants, and immigrants likely represent a large proportion of Ottawa's technology workers. Both camps would benefit from a more integrated dialogue.

---

15 See, for example, the Ottawa Talent Initiative's *Community Action Plan* of May 2004.

16 See the February 2004 report of the Internationally Trained Workers Project, the June 2004 report of the Immigrants' Employment Needs Assessment Project, and the *Trends, Opportunities and Priorities Report* of TalentWorks.

In addition, policies should be developed to ensure that foreign-born technology workers are able to apply their skills in the Canadian labour market. Immigrants in the technology sector face a double set of challenges: those related to working in the technology sector, where jobs are still relatively scarce and skill requirements are constantly evolving, and those related to being a foreign worker in Canada, such as language and cultural barriers, a lack of personal networks or Canadian job experience, and racism. Immigrants in Ottawa face even more challenges, with many jobs requiring proficiency in both English and French and Canadian citizenship.

Facilitating the integration of immigrants into the labour market has tangible economic benefits, just as allowing their skills to go unused or underused has concrete costs. A recent study by the Institute for Research on Public Policy estimated the cost to the Canadian economy of underutilising the skills of immigrants to be approximately 2 billion Cdn$ annually (Reitz, 2005, p. 3). Furthermore, skilled immigrants are highly mobile. Canada is very successful in providing skilled immigrants incentives to come to Canada, but it needs to improve the incentives to stay in the country by facilitating immigrants' economic integration. The alternative is that the permanent immigration system (along with the citizenship system) becomes a *de facto* temporary immigration system that is a costly waste of public resources.

The issue of immigrants' skills being underutilised has received a good deal of media and policy attention recently. The Ontario government has promised 12.5 million Cdn$ in the next year to assist immigrant workers in obtaining work in their field of training (Tam, 2005). In Ottawa, as mentioned previously, several community groups have been involved in bringing attention to the issue and have put forth many concrete ideas for addressing the problem. These efforts should be actively supported by government and the business community. Fields requiring accreditation, such as medicine and law, have received the most attention. However, as this chapter makes clear, foreign-born technology workers are a significant force in the Ottawa area and should be explicitly included in any strategies or policies to address the skills issue.

Finally, at the federal level, Citizenship and Immigration Canada should continue to work to develop immigration policies that meet the needs of the Canadian economy, both short-term and long-term. This should involve consultation with business leaders, including those in the technology sector. CIC has worked closely with the technology sector in the past, most notably in the establishment of the Software Pilot Project for temporary foreign workers in collaboration with SHRC. With its geographic proximity to the federal government, the Ottawa technology community can play a leading role in these efforts.

References

Beckstead, D., Brown, M., Gellatly, G., Seabornm C. (2003). *A decade of growth: the emerging geography of new economy industries in the 1990s*. The Canadian Economy in Transition Research Paper. Ottawa: Micro-economic Analysis Division, Statistics Canada.

Beckstead, D., Brown, W.M. (2005). *An anatomy of growth and decline: high-tech industries through the boom and bust years: 1997–2003*. Insights on the Canadian Economy Analytical Paper. Ottawa: Micro-economic Analysis Division, Statistics Canada.

Burt, M. (2005). *Canada's information technology and communications industry: industrial outlook. Spring 2005*. Ottawa: Conference Board of Canada.

Centre on Governance. (2002a). *Report I: Ottawa's workforce environment. Ottawa works: a mosaic of Ottawa's economic and workforce landscape*. Ottawa: Ottawa Centre for Research and Innovation.

Centre on Governance. (2002b). *Report II: profiling Ottawa's workforce. Ottawa works: a mosaic of Ottawa's economic and workforce landscape*. Ottawa: Ottawa Centre for Research and Innovation.

Citizenship and Immigration Canada (CIC). (2003a). *Immigrant occupations: recent trends and issues*. Ottawa: Citizenship and Immigration Canada.

Citizenship and Immigration Canada (CIC). (2003b). *Facts and figures 2002: immigration overview*. Ottawa: Citizenship and Immigration Canada.

Citizenship and Immigration Canada (CIC). (2004). *Facts and figures 2003: immigration overview*. Ottawa: Citizenship and Immigration Canada.

Citizenship and Immigration Canada (CIC). (2005a). *Ottawa: a comparative profile based on the 2001 census. Recent immigrants in metropolitan areas*. Ottawa: Citizenship and Immigration Canada.

Citizenship and Immigration Canada (CIC). (2005b). *Canada: a comparative profile based on the 2001 census. Recent immigrants in metropolitan areas*. Ottawa: Citizenship and Immigration Canada.

Deloitte. (2005). *Fast 500 CEO survey results*. Toronto: Deloitte.

Gagnon, R., Jacobs, L., Vaillancourt, F., Vaillancourt, L. (2003). *Canada's IT labour force, 1961–2001: four decades of growth and change*. Ottawa: Software Human Resource Council.

Green, A., Green, D. (1996). *The economic goals of Canada's immigration policy, past and present*. Research on immigration and integration in the Metropolis Working Paper Series No.96-04. Vancouver: Research on Immigration and Integration in the Metropolis.

Habtu, R. (2003). Information Technology Workers. *Perspectives on labour and income*. Catalogue No.75-001-XPE.

Hill, B. Ottawa tech firms grow by 3,100 jobs. (March 12, 2005). *Ottawa Citizen.*

Information and Communications Technologies (ICT) Branch, Industry Canada. (2004). *Canadian ICT sector regional report*. Ottawa: Industry Canada.

Information and Communications Technologies (ICT) Branch, Industry Canada. (2005). *Canadian ICT sector profile*. Ottawa: Industry Canada.

Internationally Trained Workers Project – Ottawa. (2004). *Moving forward: a strategy for the integration of internationally trained workers in Ottawa*. Ottawa: Internationally Trained Workers Project.

Jacobs, L., Vaillancourt, L., Vaillancourt, F. (2004). *The information technology labour force: Canada 2001*. Ottawa: Software Human Resource Council.

Malkam Cross-Cultural Training. (2000). *'Brain drain-brain gain' in Canadian hi tech: a focus group report*. Unpublished paper prepared for the Department of Canadian Heritage.

O'Grady, J. (2004). *The outlook for IT employment*. Presentation at the IT Human Resource Forum. Microsoft Headquarters, Mississauga, Ontario, 16 November.

Ottawa Centre for Research and Innovation. (2005). *Ottawa's tech buzz returns: employment shows steady increase*. Media Release, Ottawa, 24 January.

Ottawa Talent Initiative Project Team. (2004). *Community action plan*. Ottawa: Ottawa Talent Initiative.

Rao, B. (2001). *Economic Migrants in a Global Labour Market: A Report on the Recruitment and Retention of Asian Computer Professionals by Canadian High Tech Firms*. CPRN Discussion Paper No.W13. Ottawa: Canadian Policy Research Networks.

Reitz, J.G. (2005). *Tapping immigrants' skills: new directions for Canadian immigration policy in the knowledge economy*. IRPP Choices 11, No.1. Montréal: Institute for Research on Public Policy.

Science, Innovation and Electronic Information Division, Statistics Canada. (2001). *Information and communications technologies in Canada: a statistical profile of the ICT sector*. Ottawa: Statistics Canada.

Sutcliffe, M. (2005). 100,000 Tech jobs by '08: OCRI. January 22. *Ottawa Citizen*.

TalentWorks, Ottawa Centre for Research and Innovation. (2004a). *Immigrants' employment needs assessment project*. Ottawa: Ottawa Centre for Research and Innovation.

TalentWorks, Ottawa Centre for Research and Innovation. (2004b). *Trends, opportunities and priorities (TOP) report*. Ottawa: Ottawa Centre for Research and Innovation.

Tam, P. (2005). Ontario moves to help foreign professionals. January 26. *Ottawa Citizen*.

Wolfson, W. (2005). *Analysis of labour force survey data for the information technology occupations 2000–2004*. Ottawa: Software Human Resource Council.

Conclusion

Hugues CHICOINE and Rémy TREMBLAY

This book was about cities and knowledge, economy and innovation, planning and development, and about people and places like the Ottawa-Gatineau city-region. Economy, planning and immigrants are outstanding issues, for the authors herein identify new elements or interplays in their analyses of the country's national city-region. The result is a composite view, a loosely threaded patchwork focused on recent developments in and ideas for the local high-tech profile (or so-called clusters).

Canada is an officially bilingual country and the Ottawa-Gatineau CMA (Census Metropolitan Area), with its population of 1,130,761 in 2006, reflects that reality (Statistics Canada, 2007). However, that reality is now changing in unforeseen ways, again, after evolving from a capital city-region into a high-tech R&D nexus. In fact, almost as many people now work in the *knowledge economy* as they do for the federal government, and the region has experienced a significant increase in both its population and new companies. The book outlines the region's former and current identities, as well as setting out to qualify it as a knowledge city-region with a team made up of local stakeholders, two cities and provinces, and heavyweight central government presences. At the same time, Ottawa's balanced economy and surprisingly strong high-technology sector are not the direct result of deliberate regional planning by government.

The last decade has seen closer attention paid to economic development initiatives (although not the high-tech industry *per se*). This approach was incorporated into planning policy documents just in time for a serious decline in high-technology employment. It is, therefore, not surprising that several if not all of the players would search for more adaptable governance mechanisms in an environment where the jurisdictional neighbours, the regional co-operators, find it difficult to engage in a region-wide endeavour of collaborative governance. There are still marked differences within the region: an older manufacturing base, internal inequalities (prospering suburbs, as is the case of all CMAs in Ontario) and tensions between public sector trade unions and high-technology actors. In the 1990s, the city-region experienced massive cuts in public sector employment (called Program Review) while high-technology industries were experiencing rapid growth. Then, in 2001, the high-tech sector suffered a series of blows to its newfound confidence, with the burst of the so-called dot.com bubble. Locally, the number of employees in its high-tech establishments dropped dramatically and the crisis raised potentially anxious questions about the genuine commitments of both firms and institutional actors to address issues of personnel training and collective strategies.

Another case in point regarding infrastructure is the impact of the high-tech industry on the Ottawa-Gatineau land use-transportation system relationship with regard to cluster development. A review was conducted over a 25-year period concerning 'the messages in newspapers to local governments from the groups of people of influence' and relevant responses in official documents. Expectations and findings in the knowledge base were not reconciled whereby the claim of Ottawa as a 'world class' city or even a 'well-planned' city may hardly be justified. Clearly, the impact of the high-tech industry on the Ottawa-Gatineau land use-transportation system relationship is totally excluded from official plans and thus hinders growth. Further, the efforts expected from public authorities should be directed to assist in the recovery of a stagnant and unpredictable tech sector (renewed manufacturing processes) through joint initiatives, namely a total solution incubator focused on consumer products (ever new products) and industrial R&D including a commercialisation engine with a shift from venture financing to cooperative prototyping (new organisational styles).

Overall, the book's critical endeavour did not rely on or attempt to fit any existing knowledge city pattern, model or analytical process, whether spatial distribution, traffic, urban landscape, strategy formulation – however fragmented. There was no matching with, for example, the notion of an ideopolis, and its related methodologies concerned with identifying and measuring knowledge industries and occupations. The approach nevertheless involves a seminal and forthright endeavour into problem solving in order to make explicit the course of action. Historical and background data is used and layoffs by both government and businesses may be conceived as higher forms of negative incentives; bodies of knowledge are explored and feelings are aired which denote or call for coping or, better still, planning; finally, non-problematic statistics are computed to gain at least a limited vision of the situation at hand, a first step toward controlled development.

Other contexts still need to be processed in further investigations into Ottawa-Gatineau's creative profile in a self-reporting fashion. Of particular interest are the following directions: economic and social entrepreneurship (how sound?), planning and policy issues, and conditional instruments to overcome jurisdictional, linguistic and ethnic hurdles. Research questions should significantly address the following: (i) what local factors affect the growth in science and engineering employment, (ii) how important is the industrial make-up of a city, (iii) how important is the presence of the three universities, (iv) does a heavy governmental presence help or hinder growth, and (v) do factors that make cities attractive like amenities, cultural activities, climate, and low-crime rates also influence growth in science and engineering employment?

Reference

Statistics Canada. 2007. Community Highlights for Ottawa-Gatineau (2006). Downloaded on October 22, 2007, from http://www12.statcan.ca/english/census06/data/profiles/community/Details/Page.cfm?Lang=E&Geo1=CMA&Code1=505&Geo2=PR&Code2=35&Data=Count&SearchText=Ottawa%20%20Gatineau&SearchType=Begins& SearchPR=01&B1=All&Custom=.

Contributors

Desmond Beckstead is a Research Economist in the Micro-Economic Analysis Division at Statistics Canada. He has over 30 years of experience as a methodologist and researcher at Statistics Canada. In recent years, Mr. Beckstead's work has focused on developing measures of Canada's innovative potential using the Census, the Labour Force Survey and the National Accounts.

Amelia Brown is currently pursuing a master's degree in public affairs at Princeton University focusing on immigration policy. She holds a Van Santvoord Merle-Smith Fellowship in Public Affairs. Previously, she has worked at the Latin American Program of the Woodrow Wilson International Center for Scholars in Washington, DC and the Canada-US Fulbright Program in Ottawa. Ms. Brown has also been a Fulbright Scholar in Canada and a Junior Fellow at the International Migration Policy Program of the Carnegie Endowment for International Peace in Washington, DC. She has published several articles and co-edited two books, including *Democratic Governance and Social Inequality*, edited with Joseph S. Tulchin (Lynne Rienner Publishers, 2002). Ms. Brown holds a B.A. in international relations from Wellesley College.

Dr. W. Mark Brown is a Senior Research Economist in the Micro-Economic Analysis Division at Statistics Canada. He has a PhD in Geography from McMaster University. Dr. Brown has published numerous papers on the dynamics of change in Canada's urban and regional economies that focus on the impacts of trade, industrial restructuring, and job turnover.

After studies in linguistics, instructional media, public safety, science and technology, and cognitive sciences, *Hugues Chicoine* is at present a Télé-université of the Université du Québec à Montréal graduate student engaged in his distance education masters dissertation. A sought after research assistant, his contributions pertain mostly to geography, management and research methodologies.

Dr. Guy Chiasson is a professor at the Université du Québec en Outaouais where he teaches political science and regional development. His research interests revolve around territorial governance at the local and regional level. His latest research projects have been on forestry policy and governance, urban political reform in mid-level cities as well as regional development policies. He is an active member of the Centre de recherche en développement territorial, a Québec-wide network of researchers on regional studies.

Dr. Betsy Donald is Associate Professor in the Department of Geography at Queen's University in Kingston, Ontario. She has degrees in history, environmental studies, planning and geography from McGill, York and Toronto respectively. She teaches and does research on the urban creative economy, with recent publications in Environment and Planning A, Space & Polity, and Economic Geography. She is also a Registered Professional Planner and has consulted on a wide-range of public policy issues for all levels of government. Her most recent report was for Canadian Heritage, *Competing for Talent: Implications for Social and Cultural Policy in Canadian City-regions*. Dr. Donald currently has two SSHRC-funded research projects: one on creative class politics in Toronto and Boston, and the other on the urban creative food economy. She has received numerous awards for her research including the Governor General's Gold Medal for Academic Excellence. She is currently a Visiting Scholar at Harvard University's Rappaport Institute for Greater Boston in the Kennedy School of Government.

Dr. David Gordon is Associate Professor in the School of Urban and Regional Planning at Queen's University. He holds degrees in planning and engineering from Queen's and an MBA and doctorate from Harvard. His current research is on waterfront redevelopment and the planning history of Canada's national capital. His latest books include *Planning Twentieth Century Capital Cities* (Routledge, 2006) and *Planning Canadian Communities* (Nelson, 2007, with Gerald Hodge).

Edward T. Jackson is Chair of the Carleton Centre for Community Innovation, and Associate Professor of Public Administration and International Affairs, at Carleton University in Ottawa. Co-founder of the Community Economic Development Technical Assistance Programme and a member of Canada's National Roundtable on the Social Economy, he is a leading authority on local and regional development, advising foundations, governments, development agencies, non-profits, labour and business. His research interests include local governance and poverty reduction, knowledge management, financing civil society and citizen-directed evaluation.

Rahil Khan is a Research Associate with the Carleton Centre for Community Innovation at Carleton University. A graduate of Carleton's School of Public Policy and Administration, he has researched and written extensively on technology, sustainable development and regional development. Mr. Khan has worked as Policy Analyst with the Department of Finance and currently serves in the same capacity with Industry Canada.

John Kozuskanich is a doctoral candidate in the Department of Civil Engineering at Queen's University. He holds a B.Sc. and Master's of Urban and Regional Planning from Queen's, where he completed research on "High Tech Firm Locations and the Implications for Employment Area Planning in Ottawa."

Dr. David Large, M.Eng., M.B.A., PhD, began his professional career as an engineer, but a need to better understand the commercialisation of new technology led to a doctoral programme in Marketing at the Ivey Business School, and then to the University of Ottawa School of Management. There, he teaches, researches and consults in the fields of Marketing, Technology Marketing, and Technology-based New Enterprise. His research has been published in journals such as the Journal of Product Innovation Management and the Journal of Technology Transfer, and in several international conference proceedings such as PICMET and IEMC.

Dr. Joseph Leibovitz is a Lecturer in urban and regional studies at the Institute of Geography, The University of Edinburgh. Previously, he was a research fellow in Urban Studies at the University of Glasgow. He studied in Israel and the USA, and completed his PhD at the University of Toronto. His research interests are in comparative urban politics, and the political economy of urban and regional development. He has undertaken research and published on Scotland's urban economies, the spatial dynamics of biotechnology and old manufacturing activities in Scottish cities, the institutional foundations of learning regions in Canada, and urban regeneration in the US. More recently, his research has been focusing on the intersection of identity politics and city governance in conflict societies.

Dr. Nick Novakowski, MCIP, is an urban geographer, and an urban and environmental planner at Grenfell College, Memorial University of Newfoundland. His research interests include knowledge cities, Chinese urbanisation, and environmental planning. His consulting specialty involves economic development plus environmental assessment and, more specifically, Chinese urban development.

Dr. Rémy Tremblay received, in 2000, a PhD in Geography from the University of Ottawa. His work investigates the many sociospatial dimensions of the North American city, in both its centre core and suburbs. Between 2002 and 2004, he completed a postdoctorate transfer of work at the Institut national de la recherche scientifique (INRS-urbanisation, culture et société) in Montréal. During this two-year period, he furthered his research on cities' sociospatial form factors affecting the North American urban outlook and especially Richard Florida's creative class metatheory. Since June 2005, Dr. Tremblay has been a professor at *Télé-université* (UQAM's distance learning university). Professor Tremblay holds currently a Canada Research Chair on Knowledge Cities and director of the *Équipe de recherche sur les villes du savoir* (ERVS-research team on knowledge cities), a laboratory funded by the Canada Foundation for Innovation (CFI). He is also co-editor (in French) of the Canadian Journal of Regional Science and member of the executive committee of the *Association d'économie politique* In addition, he is a Fellow of the Royal Geographical Society (UK).

Dr. Barry Wellar, MCIP, GISP, is Distinguished Research Fellow, Transport 2000 Canada, and President, Wellar Consulting Inc. He is the author of more than 100 papers in the transportation-land use domain, and has received the Anderson Medal and the Ullman Award for his internationally recognised achievements in applied transportation research. Dr. Wellar was formerly Professor of Geography, University of Ottawa, and is a frequent media commentator on public affairs affecting the National Capital Region.

Mr. Christopher Wilson is a management consultant and the principal of Christopher Wilson & Associates, which specialises in issues of organisational development, governance, stewardship, and collaboration. Since 1997 he has been a research fellow with the University of Ottawa Centre on Governance and is a part-time lecturer in the Telfer School of Management. Mr. Wilson's main research interests involve bringing the products of research and analysis to support collaborative partnerships that foster regional innovation and community economic development. He holds an MBA from the University of Ottawa, is a member of the Institute of Public Administration Canada, the Alliance for Regional Stewardship, the National Charrette Institute, the American Chamber of Commerce Executives, and the Chaordic Commons.

"Canadian Studies": Series Titles

N° 13 – Nick NOVAKOWSKI & Rémy TREMBLAY (eds.), *Perspectives on Ottowa's High-tech Sector*, 2007, 270 p., ISBN 978-90-5201-370-1

N° 12 – Jean-François DE RAYMOND, *Diplomates écrivains du Canada. Des voix nouvelles*, 2007, 164 p., ISBN 978-90-5201-346-6

N° 11 – Claire OMHOVÈRE, *Sensing Space. The Poetics of Geography in Contemporary English-Canadian Writing*, 2007, ISBN 978-90-5201-053-3

N° 10 – Caroline ZÉAU, *L'Office national du film et le cinéma canadien*, 2006, 463 p., ISBN 978-90-5201-338-1

N° 9 – Serge JAUMAIN et Paul-André LINTEAU (dir.), *Vivre en ville. Bruxelles et Montréal aux XIXe et XXe siècles*, 2006, 375 p., ISBN 978-90-5201-334-3

N° 8 – Madeleine FRÉDÉRIC et Serge JAUMAIN (dir.), *Regards croisés sur l'histoire et la littérature acadiennes*, 2006, 195 p., ISBN 978-90-5201-333-6

N° 7 – Zilá Pierre ANCTIL & Zilá BERND (eds./dir.), *Canada from the Outside In / Le Canada vu d'ailleurs*, 2006, 294 p., ISBN 978-90-5201-041-0

N° 6 – José GOTOVITCH et Anne MORELLI (dir.), *Contester dans un pays prospère. L'extrême gauche en Belgique et au Canada*, 2007, ISBN 978-90-5201-309-1

N° 5 – Britta OLINDER (ed.), *Literary Environments. Canada and the Old World*, 2006, 246 p., ISBN 978-90-5201-296-4

N° 4 – Madeleine FRÉDÉRIC, *Polyptyque québécois. Découvrir le roman contemporain (1945-2001)*, 2005, 176 p., ISBN 978-90-5201-096-0

N° 3 – André MAGORD (dir.), *Adaptation et innovation. Expériences acadiennes contemporaines*, 2006, 274 p., ISBN 978-90-5201-072-4

N° 2 – Robert C. THOMSEN and Nanette L. HALE (eds.), *Canadian Environments. Essays in Culture, Politics and History*, 2005, 316 p., ISBN 978-90-5201-295-7

N° 1 – Serge JAUMAIN & Éric REMACLE (dir.), *Mémoire de guerre et construction de la paix. Mentalités et choix politiques. Belgique – Europe – Canada*, 2006, 316 p., ISBN 978-90-5201-266-7